Y0-AAV-888

CONCRETE STRUCTURES: REPAIR, WATERPROOFING AND PROTECTION

CONCRETE STRUCTURES: REPAIR, WATERPROOFING AND PROTECTION

PHILIP H. PERKINS
CEng, FIMunE, FASCE, FIArb, FIPHE, MIWES

A HALSTED PRESS BOOK

JOHN WILEY & SONS
NEW YORK—TORONTO

PUBLISHED IN THE U.S.A. AND CANADA BY
HALSTED PRESS
A DIVISION OF JOHN WILEY & SONS, INC., NEW YORK

Library of Congress Cataloging in Publication Data

Perkins, Philip Harold.
Concrete structures.

"A Halsted Press book."
Includes index.
1. Concrete construction—Maintenance and repair. 2. Waterproofing. 3. Concrete coatings. I. Title.
TA681.P42 624'.1834'028 77–23140
ISBN 0–470–99087–2

WITH 76 ILLUSTRATIONS

Printed in Great Britain by Galliard (Printers) Limited, Great Yarmouth

Preface

While there are examples of reinforced concrete structures built towards the end of the nineteenth century and the early part of the twentieth, concrete as a general constructional material only began to be used on a larger scale after the end of the First World War. The military requirements of the Second World War and the large development and rebuilding programme which followed it, established concrete as the major constructional material. Structures erected in the 1920s and 1930s are now between 40 and 50 years old, which will be within the memory of many readers of this book.

The use of a relatively new building material inevitably brings problems which were not anticipated initially and disappointments are common. Reinforced concrete is no exception and due to the increasing age of the early structures, the need for repair and renovation is increasing.

While the structural and architectural design of buildings varies widely from one country to another, the principles of repair are more universally applicable. Therefore the author hopes that the contents of the book will be useful to a very wide range of persons who are responsible for the maintenance of concrete structures of all types.

The opinions and recommendations in this book are those of the author, but he is indebted to his colleagues in the Cement and Concrete Association and to staff in the leading firms which specialise in the repair, protection and waterproofing of all types of concrete structures. To all these people the author expresses his sincere thanks. He also wishes to thank the staff of the Cement and Concrete Associations of Australia, New Zealand and South Africa for their help in compiling the relevant sections in Appendices 1 and 3.

Contents

Introduction

There are certain basic principles which are common to the repair of all types of reinforced concrete structures. It is the objective of the author to present these principles in a concise form and then extend them to the various structures dealt with in the book.

The first signs of deterioration in reinforced concrete buildings are usually fine cracks and rust stains which may be accompanied by spalling of the concrete. The rust stains and spalling are caused by corrosion (rusting) of the reinforcement, while the cracking may be due to other factors.

The same sequence of events can occur in marine structures except where there is physical damage to the concrete caused directly or indirectly by wave action. With certain other structures however, there may be chemical attack on the concrete itself, and in extreme cases this can result in rapid deterioration.

The rust stains and spalling of concrete caused by corrosion of the reinforcement occur before attack on the steel has progressed very far. This is because corrosion products (mostly oxides of iron) occupy a larger volume than the original steel and the expansion which accompanies the formation of the rust causes cracking and spalling of the concrete cover to the reinforcement. This means that unless there is serious neglect, the trouble is detected at an early stage and long before deterioration has progressed far enough to endanger the stability of the structure.

The first step to be taken when called to investigate a concrete structure which is in need of repair is to establish the cause of the deterioration. The experience of the author is that in the vast majority of cases there is no danger of early collapse, and this is not likely to occur unless there is a long delay in the execution of the necessary repairs.

The usual causes of deterioration are inadequate cover to the reinforcement, poor quality concrete, thermal contraction and/or shrinkage cracks which have been allowed to remain without repair. These defects are, in a few cases, aggravated by the presence of calcium chloride in the concrete, the chloride ions being a contributory factor in the corrosion of the reinforcement. The existence of chlorides in the concrete can in certain circumstances seriously reduce the durability of any repair work.

The term 'concrete structures' used throughout this book refers to concrete made with Portland type cements. In certain cases recommendations have been made for the use of high alumina cement for repair work. This may cause some surprise in view of the fact that high alumina cement concrete is no longer a 'deemed to satisfy' material for the purpose of Part D of the Building Regulations; also, all reference to this cement (HAC), has been omitted from Building Research Establishment Digest No. 174—Concrete in Sulphate Bearing Soils and Ground Water.

However, the author feels that the use of high alumina cement for specific types of repair work can give entirely satisfactory results. He wonders whether the changes in the Building Regulations and in official publications might have been over hasty. He hopes that as a result of the testing and research now being carried out, some modification to the present embargo will emerge.

CHAPTER 1

The Principal Materials Used in the Repair of All Types of Concrete Structures

The range of materials which can be effectively used for the durable repair of concrete structures is fairly limited. Those most widely used are concrete and mortar, made as far as practicable with the same type of cement and aggregates as were used in the original structure. When deterioration is due to chemical attack, it may be necessary to use a different cement and/or protective coatings.

When repaired areas fail it is in many cases due to failure or partial failure of the bond between the old and new work. The standard of bond developed between the old and new concrete is directly related to the care taken in the preparation of the base concrete. In recent years a great deal of attention has been paid to the development of bonding agents.

The vast majority of concrete structures which need repair are reinforced and the corrosion of the reinforcement plays an important part in the deterioration of the structure. When ferrous metals corrode, the corrosion products occupy a larger volume than the original metal and the resulting expansion causes disintegration (spalling) of part of the surrounding concrete.

The steel passivating properties of the original concrete are of great importance in providing long term durability to the reinforcement and to the structure as a whole. Good quality Portland cement concrete and mortar can provide permanent protection to the steel due largely to the alkaline environment created by the cement paste. This important matter is dealt with in some detail in Chapter 2.

The selection of suitable materials to replace defective and spalled concrete and to reintroduce a protective and durable environment around the reinforcement is of great importance.

1.1 CEMENTS

For the purpose of this book cements are classified as Portland cements and non-Portland cements.

The quantity of Portland cement used in the construction industry in the UK far exceeds that of all other types and in 1973 amounted to about 20 million tonnes.

1.1.1 Portland Cement: Ordinary and Rapid Hardening

In the UK these two cements are covered by British Standard BS 12 and form the bulk of the Portland cements used for the repair of concrete structures, with ordinary Portland predominating.

The basic difference between the two cements is the rate of gain of strength. The increase with the rapid hardening cement is largely due to the finer grinding and the cement usually has a specific surface of about 4300 cm^2/gm. The rapid hardening is accompanied by an increase in the rate of evolution of heat of hydration which in turn raises the temperature of the maturing concrete during the first 15 to 40 h after casting. It should be noted however, that the rate of hardening and evolution of heat of hydration does not depend on the fineness of grinding alone, and that the chemical composition of the cement also plays a part.

1.1.2 Sulphate-Resisting Portland Cement

The relevant British Standard is BS 4027 and the cement is similar in its strength and other physical properties to ordinary Portland cement. However, it is generally darker in colour than most ordinary and rapid hardening Portland cements. The essential difference is in the limitation of the tricalcium aluminate (C_3A) content to a maximum of 3%.

It is the tricalcium aluminate in Portland cement which is attacked by sulphates in solution and this chemical reaction results in the formation of ettringite which can have a disruptive effect on the concrete, causing dimensional changes and reduction in strength. It is advisable to consult the cement manufacturer before using any type of admixture with this type of cement, but calcium chloride should not be used as the sulphate resistance in the long term will be reduced.

Sulphate-resisting Portland cement, in common with all Portland cements is vulnerable to acid attack.

1.1.3 White and Coloured Portland Cements

White Portland cement is a true Portland cement and complies with BS 12 (Portland cement: ordinary and rapid hardening). The special point about this cement is that the raw materials are specially selected; the clay is a white china clay and the manganese and iron content is kept to an absolute minimum.

Coloured Portland cements, other than white and the pastel shades, generally consist of ordinary Portland with a pigment ground in at the works. The pigments used are covered by BS 1014—Pigments for cement and concrete. Coloured Portland cements are now included in BS 12.

These cements would only be used in cases where a colour match with the existing concrete was desirable. However, due to colour changes with age and weathering, it is likely that the best results would be obtained by using pigments and trial mixes.

A question is sometimes raised as to whether it is necessary to increase the cement content of a concrete or mortar mix to allow for the presence of the pigment in the cement. It is not possible to give an answer to this in the form of a straight 'yes' or 'no'. Where durability and/or impermeability are important factors, the author considers that allowance for the pigment should be made by increasing the specified cement content by weight of pigment. Where strength is overriding, then the only satisfactory solution is to make trial mixes because the strength of ordinary Portland cement varies within certain limits.

At the time of writing this book, coloured (pigmented) Portland cement was no longer available in small quantities on the UK market.

1.1.4 Ultra-High Early Strength Portland Cement

This is a cement of the true Portland type which appeared on the UK market a few years ago under the trade name of 'Swiftcrete'. At the time of writing there is no British Standard for this cement, but an Agrément Certificate, No. 73/170 has been issued. It is described by the manufacturers as an extremely finely ground Portland cement which contains a higher proportion of gypsum than ordinary Portland, but otherwise complies with BS 12; it contains no other additives.

The specific surface of the cement is between 7000 and 9000 cm^2/gm compared with an average ordinary Portland of 3400 cm^2/gm, and 4300 cm^2/gm for an average rapid hardening Portland.

The tests carried out by the Agrément Board confirm that the strength at 24 h is not less than the strength of a similar mix of concrete made with rapid hardening Portland cement at seven days. The Certificate makes a number of statements relating to the use of this cement and includes a table giving a detailed comparison between 'Swiftcrete' and ordinary and rapid hardening Portland cements.

This cement is about 50–60% more expensive than ordinary Portland, but this is not significant in those special cases where very high early strength is required.

1.1.5 High Alumina Cement (HAC)

While this book was being written a great controversy arose in the UK over the use of HAC concrete in building structures. This was caused by the unfortunate failure of two roof beams in a school. A considerable amount of ill-informed comment was made in the national press and in some technical journals. Because of this the author considers it desirable to present a brief and balanced appraisal of HAC concrete and to show how this differs in important respects from Portland cement concrete.

High alumina cement is covered by BS 915. About 60% of the world's requirement of this cement (outside the USSR) is produced by Lafarge Fondu International; of this, about 90% is used for refractory concrete where structural strength may be of little significance.

HAC differs fundamentally in chemical composition from Portland cement as it consists predominantly of calcium aluminates. It is a much darker colour (dark grey) than ordinary and rapid hardening Portland cements. The lighter shades of HAC and darker shades of sulphate-resisting Portland cement may approach each other in colour. The cement sets and hardens when mixed with water, and under normal conditions of temperature the setting time is similar to that of ordinary Portland cement. However, with increase in ambient temperature, under normal climatic conditions existing in the UK, the setting time tends to increase, whereas the reverse is the case with Portland cement. In countries where ambient temperatures are higher it should be noted that with temperatures above approx. 30 °C, the setting time is reduced and the set becomes increasingly faster and in extreme cases may even 'flash' set. On the other hand, the rate of gain of strength of HAC is very rapid, normally reaching about 80% of its strength in 24 h, compared with Portland cement which reaches about 80–85% of its ultimate strength in 28 days. The rapid increase in strength is accompanied by a rapid evolution of the heat of hydration. This has advantages and disadvantages. It is extremely useful when working in low temperatures; also it enables emergency repairs and similar work to be carried out within a short period of time. Wet curing is essential to prevent premature drying out of the concrete surface, and the concrete must be placed in relatively thin layers so that the heat can be dissipated quickly. It is usually recommended that the thickness of 'lifts' should be restricted so that no joint is more than 500 mm from a cured surface. In practical terms, this means that a wall 1 m thick can be cast continuously without horizontal joints (as would be done with Portland cement concrete). However in certain cases (such as furnace foundations) it is necessary to cast large volumes of concrete made with HAC. For such structures special mesh reinforcement and cooling tubes carrying cold or ice-water are cast into the foundation block to control thermal cracking. Good curing techniques are also necessary.

HAC concrete which has been correctly proportioned, placed and cured, exhibits improved resistance compared with Portland cement concrete, to many chemical compounds, including sulphates, sugars, vegetable oils and certain dilute acids including lactic acid. HAC is not classified specifically as a chemically resistant cement, and chemicals including chlorides and dilute caustic and non-caustic alkalis, which have little effect on Portland cement concrete, will attack HAC concrete.

To achieve maximum chemical resistance and long term durability, the same basic principles used for Portland cement concrete should be adopted. These are, high cement content not less than 400 kg/m^3, low water/cement ratio not higher than 0·40, thorough compaction, and careful water curing. This gives a dense, impermeable concrete which is essential for durability. The long term durability of HAC concrete in sea water is referred to in Chapter 5.

HAC is more coarsely ground than ordinary Portland cement (specific surface about 3000 cm^2/gm, compared with about 3500 cm^2/gm for an average OPC). This, together with the physical characteristics of the cement particles, enables workable mixes to be prepared with lower free water/cement ratios than mixes with Portland cement having the same workability.

It is at this point that mention must be made of the most controversial matter relating to HAC, namely, 'conversion'. A great deal has been written about conversion, but briefly, the hydrates which cause HAC to develop strength rapidly are metastable and convert into a denser and more stable form. Unpublished work by D. E. Shirley, and a report by French and coworkers in the journal 'Concrete' (August 1971, pp. 3–8), and a report by Dr C. M. George in 1975, suggested that the phenomenon of conversion could in fact be allowed for in a relatively simple manner in the mix design procedure. In short, the effect of conversion depends largely on the rate at which it takes place, and this in turn is determined largely by the temperature during the maturing period and of the environment in which the concrete exists during its working life. The original water/cement (w/c) ratio will basically determine the strength of the concrete after conversion. There is evidence to suggest that conversion may be of little practical significance provided that the concrete mix was originally designed for full compaction at a w/c ratio appropriate to the maximum sustained temperature during the life of the structure. An acceptable formula incorporating w/c and temperature and relating these to minimum strength, has, unfortunately, not been generally agreed. The nearest to this, was contained in an article in the 'New Civil Engineer' (21 March 1974, pp. 40 and 43) and the information given in the publication written by Dr C. M. George of Lafarge Fondu International.

A further point in connection with conversion is that after the minimum point has been reached, a high quality concrete with a low w/c ratio will start to increase in strength.

Its use in repairs is likely to be confined to:

(a) Structures in which the original concrete was made with HAC.
(b) Emergency repairs to Portland cement concrete floors.
(c) The floors of cold stores.
(d) Repairs to marine structures.

(e) Gunite or rendering in cases where the chemical and/or heat resistance of the cement is an advantage compared with Portland cement.

Following the unfortunate 'panic' mentioned at the beginning of this section, considerable efforts were made to develop a relatively quick, simple and reliable test to determine on site, whether a particular concrete member contains HAC. The Building Research Station largely succeeded in this and the following is an extract from their Information Sheet IS 15/74 of October 1974:

> 'The test involves extraction of a 1 gm sample of powdered concrete with 10 ml of 0·1 N solution of sodium hydroxide and, after filtration, treating the extracted solution with "oxine" (8-hydroxyquinoline) under suitable conditions. Abundant formation of a yellow precipitate indicates the presence of high alumina cement in the concrete.
>
> The yellow precipitate indicates that the concrete contains an appreciable quantity of aluminium. When the test is applied to concretes made with Portland type cements, the final solution remains reasonably clear or slightly cloudy.'

1.1.6 Chemically Resistant Cements

These special cements are not used for concrete except where very small quantities are required, but are used for mortar for bedding and jointing chemically resistant tiles and bricks. This type of construction is used in lining tanks which hold very aggressive liquids.

The two basic types of cement are resin cements and silicate cements; certain grades of the latter are resistant to high temperatures. Cements can now be obtained which are resistant to most chemicals in common use with the general exception of hydrofluoric acid in concentrations above 40%.

These cements usually consist of a powder and a special gauging liquid (sometimes called a syrup) which are mixed together in the prescribed proportions.

1.2 COMPARISON OF UK AND US PORTLAND CEMENTS

In view of the employment of consultants and contractors from both the UK and USA on international contracts, it is considered that some general information on Portland type cements from these two countries would be useful.

The direct comparison of standard specifications of one country with those of another, can be very misleading. The requirements are different, and the methods of test are different. However, Table 1.1 is intended to show approximate comparisons. The most that can be said is that for conditions where a sulphate-resisting Portland cement to BS 4027 would be specified, it is likely that a US Portland cement to C 150-67 Type V, would also be satisfactory.

TABLE 1.1

Nature of cement	ASTM No. & Type	British Standard No.
Ordinary Portland	C 150-67 Type I	BS 12:1958
Sulphate-resisting/low heat	C 150-67 Type II	No equivalent material
Rapid hardening	C 150-67 Type III	BS 12:1958
Low heat	C 150-67 Type IV	BS 1370:1958
Sulphate-resisting	C 150-67 Type V	BS 4027:1966
Air-entraining (various)	C 175-67 Types IA, IIA, IIIA	No equivalent materials

1.3 STEEL REINFORCEMENT

Steel reinforcement for concrete including prestressed concrete is covered by seven British Standards which are listed in Appendix 1.

For normal reinforced concrete there are three types of reinforcing steel:

Mild steel plain hot rolled bars (BS 4449)
Cold worked steel bars (BS 4461)
Steel fabric (BS 4483)

High tensile steel is identified in various ways; there are cold worked twisted bars where the twisting is clearly seen, cold worked ribbed bars, and hot rolled deformed bars.

1.3.1 GALVANISED REINFORCEMENT

At the time of writing this book (1975), there is no British Standard for galvanised steel reinforcement nor for galvanised prestressing wire although both are used to a limited extent in the UK and to a greater extent in the USA. The advocates of the more extensive use of galvanised reinforcement make the point that the whole structural stability and long term durability of a reinforced concrete structure depend on the assumption that the reinforcement will not corrode to any significant degree during the lifetime of the structure. Therefore any practical measures to help ensure this are desirable. It is a fact that steel is more vulnerable to corrosion than concrete. On the other hand, it is important to keep a sense of proportion and to realise that when steel is encased in dense good quality concrete with adequate cover, corrosion of the reinforcement does not occur unless external conditions are particularly aggressive, in which case, the concrete itself is attacked first, or physical damage occurs to the concrete so as to expose the steel.

However, conditions sometimes arise when the use of galvanised reinforcement and prestressing wire is justified and some of these circumstances are mentioned later in this book.

Further information on the use etc. of galvanised reinforcement is given in Chapter 2. Galvanising consists of coating the steel with zinc, either by dipping the steel into tanks of molten zinc, or by electro-deposition from an aqueous solution.

1.3.2 Stainless Steel

There are three basic groups of stainless steel: martensitic, ferritic, and austenitic. Of these, it is the austenitic steels which are the most widely used in building and engineering and the information which follows relates to this group. Austenitic steel is an alloy of iron, chromium and nickel, and two types in this group contain also a small percentage of molybdenum. The type most generally suitable for use in external repair work is En58J (also known as 316 steel), which contains 18% chromium, 10% nickel and 3% molybdenum. It is very resistant to corrosion but is also very expensive compared with mild steel. The steel can be welded easily. However, it may be non-magnetic or only slightly magnetic, and therefore it may not be possible to locate it with a normal type of cover meter. A further point is that mild steel in contact with stainless steel may suffer accelerated corrosion, *i.e.* it may be anodic to stainless steel. It is recommended that advice be sought from one of the organisations listed in Appendix 3.

This steel (En58J) is covered by BS 1449: Part 4, for flat rolled strip, while bars are covered by BS 970: Part 4.

The use of stainless steel is only justified in exceptional cases where protection of normal steel cannot be relied upon due to difficulty of application or difficulty of maintenance of the protective treatment. In repair work its use is likely to be confined to pins, fixings, and anchors, and stainless steel mesh for reinforcing relatively small areas which are repaired with mortar.

The cost of stainless steel is much higher than mild steel by a factor of about eight.

1.4 NON-FERROUS METALS

A limited number of non-ferrous metals are used in repair work; when these are used, it is hoped that the following information and recommendations will prove useful.

1.4.1 ALUMINIUM

If unanodised aluminium is used in direct contact with damp concrete the metal should be protected by a thick coating of bituminous paint or other material which will not be attacked by the caustic alkalis in the cement.

Anodising improves the durability, resistance to corrosion and the aesthetic appearance of aluminium. Anodising consists of the imposition on the metal of a thick coating of the metallic oxide, built up in layers by a special process. The oxides can be pigmented. Aluminium oxide is inert and relatively impermeable, particularly if the surface is sealed by immersion in hot water containing chromates.

Generally aluminium is anodic to steel, but certain alloys of aluminium are cathodic to steel. Therefore, care must be exercised if the two metals are in contact.

1.4.2 COPPER

Copper is resistant to attack by most conditions met in building construction and water and sewage works. It is not corroded by Portland cement concrete unless chlorides are present in the concrete. Copper is cathodic to steel and therefore if the two metals are in contact in the presence of moisture, there is danger that the steel will corrode and the copper will be protected.

1.4.3 PHOSPHOR-BRONZE AND GUNMETAL

Bronze is a copper–tin alloy, and phosphor-bronze contains phosphorus as copper phosphide. It is resistant to conditions which would result in corrosion of ferrous metals. The same general remarks apply to gunmetal which is bronze with about 9% tin.

1.5 AGGREGATES

For repairs, aggregates for concrete and mortar are generally considered less important than for new structures. Nevertheless, it is felt that some general remarks may be useful, as some shortcomings in the quality of the aggregates may be partly responsible for the deterioration of the concrete in certain areas. In the UK, aggregates obtained from 'natural sources', such as gravels, pit sand, quarried rock and sea-dredged material, should comply with BS 882—Aggregates from natural sources for concrete. Tests on aggregates are covered by BS 812—Methods for sampling and testing of mineral aggregates, sands and fillers. It is important to note that the British Standard 882, contains the following statement:

> 'No simple tests for the durability and frost resistance of concrete or for corrosion of reinforcement can be applied and experience of the

properties of concrete made with the type of aggregates in question and a knowledge of their source are the only reliable means of assessment....'

Shrinkage of natural aggregates, when excessive, can create problems. In the UK this characteristic has so far only appeared in certain aggregates in Scotland, but in other parts of the world it has proved to be a significant factor in the selection of aggregate sources. This characteristic is dealt with in detail in the Building Research Establishment's Digest No. 35—Shrinkage of Natural Aggregates in Concrete. Aggregate–alkali reaction is another matter which should be considered when dealing with aggregates from a previously unknown or untried source. This particular problem is extremely complicated and is outside the scope of this book.

Matters which can be important, depending on the circumstances of each case, include the following:

(a) Source and classification.
(b) Particle shape and surface texture.
(c) Grading (sieve analysis) including clay, silt and dust content.
(d) Organic impurities.
(e) Salt content, particularly of the fine aggregate, with particular reference to chlorides and sulphates.
(f) Mechanical properties.
(g) Flakiness index.
(h) Shell content; this applies principally to sea-dredged aggregates.

There is an increasing tendency in the UK to view with concern any significant concentration of chlorides in reinforced concrete. At the time this book was being written, Code of Practice, CP.110—The Structural Use of Concrete, clause 6.2.4.2, allowed a maximum anhydrous calcium chloride content of 1·5% by weight of cement, in reinforced concrete, with complete prohibition in prestressed concrete. However, moves were afoot to introduce a revision which would effectively ensure 'chloride free' concrete. In practice this would mean a chloride ion concentration not exceeding that found in normal drinking water. The use of sea water for mixing concrete is discussed in Chapter 2.

It has been found particularly difficult to place limits and definitions on organic impurities. The most satisfactory procedure is to carry out tests using the contaminated aggregates and 'pure' aggregate as a control, and see what changes occur in the setting time, rate of gain of strength, and 7 and 28 day strengths between the two groups.

There are no national regulations for shell content of aggregates, but the following can be considered as reasonable for most purposes:

Aggregate size (mm)	Maximum shell content (% of dry weight)
40	2
20	5
10	15
fine aggregate	30

The effect of the shell content depends on the size of the shells and their shape and the environmental conditions (exposure) under which the concrete will have to function. Large shells can adversely affect durability, and can be detrimental to the appearance of fair-faced concrete. Poorly shaped fine shell can increase the water demand of the mix. As shell is mostly calcareous, this must be allowed for in chemical analysis of concrete to determine cement content, in the same way as for limestone aggregate.

1.6 ADMIXTURES FOR CONCRETE

A simple definition of an admixture is that it is a chemical compound which is added to concrete, mortar or grout at the time of mixing for the purpose of imparting some additional and desirable characteristic(s) to the mix.

Admixtures are sometimes referred to as 'additives', but it is better to use the latter word for the addition of chemical compounds to cement at the cement works, where they are ground-in at the time of manufacture.

Admixtures should only be used when they are really required to produce a particular result which cannot be obtained by normal mix design.

Admixtures should not be used with any cement except ordinary and rapid hardening Portland cement without the approval of the cement manufacturer.

The following are the main purposes for which admixtures are used:

(a) To accelerate the setting of the cement and the hardening of the concrete, mortar or grout; these compounds are known as accelerators.
(b) To retard the setting of the cement and slow down the rate of hardening of the mix. These are known as retarders.
(c) To entrain air in the mix (air-entraining agents). These compounds give an air-entrained mix, which should not be confused with an aerated mix. The latter is obtained by quite different compounds and is used for different purposes.
(d) To entrain air in the mix (air-entraining agents). These compounds included in this category are water-reducing admixtures and 'waterproofers'.

1.6.1 ACCELERATORS

Accelerators can be useful in cold weather and urgent repair work, such as work between the tides and patching of floors.

The great majority of accelerators used in concrete are based on calcium chloride ($CaCl_2$) as the active ingredient. The use of this compound, apart from speeding up the chemical reaction of cement and water, has certain other effects, the most important of which are:

(a) Calcium chloride is very aggressive to ferrous metals.
(b) It increases drying shrinkage of the mix.
(c) It reduces the sulphate resistance of sulphate-resisting Portland cement.

An impartial technical appraisal would show that $CaCl_2$ can be used safely in concrete provided a long list of conditions are completely satisfied. In practice on construction sites it is often not possible to ensure 100% compliance with these conditions. Therefore the author recommends that calcium chloride should not be used except for temporary, emergency work, unless the professional man responsible for the supervision of the work is entirely satisfied that all necessary conditions can be met consistently.

Because of the serious disadvantages of calcium chloride mentioned above, considerable efforts have been made to find a satisfactory substitute as the basis for accelerators. So far only two compounds have met with even a limited degree of practical success, and these are calcium formate and sodium carbonate.

The most effective and satisfactory method of speeding up the setting and hardening of Portland cement concrete is the use of heated concrete or the application of heat to the concrete after casting.

1.6.2 RETARDERS

There are two main uses for retarders. One is as an integral part of a concrete mix when it is required to extend the setting time of the cement and reduce the rate of hardening of the concrete. The other is when the retarder is used on formwork to retard the setting and hardening of the surface only of the concrete, in order to facilitate work on the concrete surface when the formwork is removed.

Retarders are usually sugars and similar compounds, but borax is also used. The reaction between retarders and Portland cement is a very complicated one. It is affected by the chemical composition of the cement and the temperature of the concrete as it is maturing. Therefore the period of retardation can only be estimated approximately and accurate reproduction of results is impractical.

1.6.3 AIR-ENTRAINING AGENTS

Air-entrained concrete is used for roads and external pavings, to resist the disintegrating effects of frost and de-icing salts. Air-entraining agents, however, can be useful for concrete repair work and for mortar used on repairs to structures on very exposed sites in the Northern parts of the UK, as well as for marine structures in Northern latitudes.

While these compounds were originally introduced to provide a concrete which would resist frost action, it has been found that they impart other beneficial characteristics to the mix. These are:

(a) They help to reduce, and may, in favourable circumstances, completely eliminate plastic cracking.
(b) They help to reduce a tendency to water scour on the surface of fair-faced concrete.
(c) They improve workability.

The best air-entraining agents are resins. They must be carefully dispensed into the mixing water, so that the dosage is accurately controlled and the compound is uniformly distributed throughout each batch. Their effect is to produce a large quantity of minute bubbles of air which alter the pore structure of the concrete. The effect of this entrainment of air (about $4\frac{1}{2}\% \pm 1\frac{1}{2}\%$) is to slightly reduce the compressive strength of the concrete but at the same time to improve the workability.

Air-entraining agents should not be confused with such compounds as aluminium powder which is used for the production of aerated lightweight concrete and mortar.

1.6.4 Plasticisers or Workability Aids

These admixtures can be divided into two main types:

Lignosulphonates, also known as lignins, and soaps or stearates
Finely divided powders

The lignosulphonates and stearates act very largely as lubricants and in this way the amount of water required in the mix to obtain a predetermined workability can be reduced; or, for a given w/c ratio the workability is increased. Some of these compounds, namely the stearates, impart a degree of water repellency to the concrete, mortar or grout and these are sometimes referred to as waterproofers. An overdose of certain of these compounds will produce a set-retarding effect; serious overdosing can result in a permanent reduction in compressive strength.

The finely divided powders include pulverised fuel ash (PFA), powdered hydrated lime, powdered limestone and bentonite. Portland cement itself is a good plasticiser and an increase in cement content may help to overcome problems of segregation and harshness. Depending on the characteristics of the powder and quantity used, the water demand of the mix may be increased but its cohesiveness may be improved.

To sum up, both types (lignins and stearates, and powders) have their specific uses and both can be considered as reliable provided they are correctly used. They help to achieve a workable and cohesive mix which can be compacted under the action of poker vibrators and vibrating beams.

It should be remembered that many manufacturers of admixtures produce compounds which serve more than one purpose; this means that one can obtain plasticisers which also act as retarders, while other types act as accelerators. It is therefore important to obtain complete information on the basic composition and all the effects of a particular admixture before deciding to use it.

1.6.5 Superplasticising Admixtures for Concrete

These new materials are a relatively new type of chemical admixture as far as their use in the UK is concerned. They have however, been in commercial use in Japan since about 1967 and in Germany since about 1972. It is important to appreciate that superplasticisers are chemically distinct from the normal workability aids at present in general use; they can be used with confidence at high dosage levels, subject to the general conditions set out in this section.

Superplasticisers can be used for two purposes:

1. To produce a concrete having a virtually collapse slump, *i.e.* a 'flowable' concrete.
2. To produce a concrete with normal workability but with a very low w/c ratio, resulting in a high strength concrete.

It is not the intention in this section to discuss the chemical composition of the various types of superplasticisers available in the UK, but this can be summarised as follows:

Group 1—sulphonated melamine formaldehyde
Group 2—naphthalene sulphonated formaldehyde
Group 3—modified ligno-sulphonates

Of these, the first two groups appear to be the more reliable and effective.

Research work and site tests in Germany, Japan and the UK have shown that so far the use of these compounds has not revealed any adverse effects on the durability of the concrete, nor on the ability of the concrete to passivate steel reinforcement and protect it from corrosion, nor has any reduction in the strength of the concrete over a long period of time been detected.

In order to obtain concrete which will 'flow' but not segregate, it is necessary to start with a slump of about 50–75 mm before the superplasticiser is added. The fine aggregate (sand) content of the mix must be increased by about 4–5% and the coarse aggregate correspondingly reduced so that the aggregate/cement ratio remains unchanged. The superplasticiser is added to the mix after the addition of the water and then mixing should be continued for at least another two minutes.

Correctly made superplasticised concrete has a slump of about 200 mm (when it is required to be flowable), and is almost self-levelling; it is cohesive and will not segregate. Strict site control of the mix proportions, especially the sand content and the original slump, are essential if segregation is to be avoided.

Because of the danger of segregation when using these admixtures, it is most important that trial mixes be carried out prior to their use and to ensure that the final mix design, including the type and grading of the fine and coarse aggregates is exactly followed when the concrete is made.

It should be noted that maximum workability is usually only retained for a period of about 30–60 min, and then the concrete rapidly reverts to normal conditions of slump. The information so far available on superplasticisers shows that they are a very useful addition to the concreting industry. One of the basic principles of good quality concrete in the structure is that it must be thoroughly compacted. Full compaction can often prove difficult and in some cases almost impossible to achieve if high strength is to be maintained. This is particularly so when repairs with concrete have to be carried out to walls and columns as well as members of small section.

For further information on superplasticisers the reader is referred to the Bibliography at the end of this chapter.

1.6.6 Pulverised Fuel Ash (PFA)

PFA is produced in very large quantities as the residue from coal burning electricity generating stations. It is a fine powder, having a specific surface similar to ordinary Portland cement, *i.e.* about 3400 cm^2/gm. The specific gravity however is appreciably lower than that of cement, being in the range of 1·9 to 2·3 (OPC is about 3·12). The main chemical compounds in PFA are oxides of silicon, iron and aluminium together with some carbon and sulphur. The relevant British Standard is BS 3892: Pulverised Fuel Ash for Use in Concrete. The Standard limits the amount of combustible material and sulphur compounds. PFA possesses some pozzolanic properties and this is the principal reason why it is used in certain classes of concrete and grout.

A proposal to use PFA in structural concrete should take the following factors into consideration:

(a) PFA should only be incorporated into concrete, grout or mortar with the written approval of the client's technical representative.
(b) For all structural concrete, a minimum cement content should be specified for reasons of durability, and the replacement of cement by PFA should not reduce the cement content to below this minimum figure.
(c) The chemical composition and the pozzolanic properties of PFA

are closely related to the coal from which it is derived and the combustion characteristics of the power station where the coal is burnt. These variations can affect the compressive strength of the concrete, both at early ages and at later periods such as 3, 6, and 12 months.

Consideration of item (b) above will show that this is more important than may appear at first sight. Minimum cement contents are based on experience over many years and these have a direct influence on durability. Durability is a very wide term and includes resistance to such factors as freezing and thawing, moisture penetration (impermeability), chemical attack, and depth of carbonation in relation to time. The author has seen no published reports from independent authorities which show that concrete in relatively thin structured members in which part of the minimum cement content has been replaced by PFA, is equally as durable as concrete in which such replacement has not taken place.

Consideration must also be given to the effect of PFA on the protection of reinforcement. One of the most important factors in the protection of steel by concrete against corrosion is the intense alkalinity of the Portland cement matrix which has a pH of about 11·5. PFA is only slightly alkaline and has a pH close to the neutral point of 7·0. Therefore any dilution of the matrix which could reduce its alkalinity, may have adverse long term effects on the protection of the reinforcement.

The author does not consider that the addition of PFA to ordinary Portland cement is an effective substitute for sulphate-resisting Portland cement when the latter is required to resist sulphate attack. One of the factors affecting the effect on sulphate resistance of an OPC/PFA blended cement would be the C_3A (tricalcium aluminate) content of the cement. A further practical point when considering the addition of PFA on site to Portland cement concrete mix, is that this introduces one more constituent which has to be carefully controlled.

1.7 JOINT FILLERS AND SEALANTS

1.7.1 Fillers

Fillers for joints are sometimes known as 'backup' materials. They are used in full movement joints to provide a base for the sealant and also to prevent the ingress during the construction period of stones and debris which may prevent the joint from closing. Materials for these fillers include specially prepared fibres, cellular rubber and granulated cork compounds. The material used should fulfil the following requirements:

(a) It must be very durable.
(b) It must be chemically inert.

(c) When in contact with potable water it must be non-toxic and non-tainting and should not support bacterial or fungoid growth.
(d) It must be resilient and should not extrude so as to interfere with the sealing compound, and should not bond to the sealant as this latter could induce undesirable stress in the sealant.
(e) It should be easily formed to the correct dimensions and be readily inserted into the joint.

Materials are now available which fulfil the above conditions satisfactorily.

1.7.2 SEALANTS

The materials used to seal joints in structures can be conveniently divided into two basic groups:

(a) Preformed materials
(b) Insitu compounds

To be satisfactory, both groups should possess the following characteristics:

(i) For external use, or use in liquid retaining structures, the sealant itself must be impermeable.
(ii) It must be very durable as periodic renewal may be difficult and expensive. Ideally, its life should be the same as the structure of which it forms a part; this condition is not fulfilled by any known sealant at the present time.
(iii) It must bond to the sides of the groove in which it is inserted. In practical terms this means that the sealant should bond well to damp concrete.
(iv) In potable water tanks the material must be non-toxic and non-tainting and should not support the growth of bacteria and fungi etc.
(v) As the joint opens and closes, the sealant must deform in response to that movement without undergoing any change which will adversely affect its integrity.
(vi) It should be comparatively easy to install under the weather and site conditions relevant to the location of the structure.

The sealant, whether preformed or insitu, is normally accommodated in a groove in the concrete. Research in both the UK and the USA has shown that the shape and dimensions of the groove are important in ensuring a satisfactory and durable seal. For detailed information on this subject reference should be made to the Bibliography at the end of this chapter. A simplified but reasonably correct statement would be to the effect that the depth of the groove which is filled by the sealant should be approximately half the width.

A preformed channel-shaped gasket made of neoprene has come onto the market. This spans the joint and is fixed with an adhesive into narrow grooves cut parallel to each side of the joint. Figure 4.20 (p. 165) shows this gasket.

Preformed Materials

Preformed joint sealants at present take up only a small percentage of the market, but this share is steadily increasing. The majority of the high grade material is based on neoprene and is imported from the Continent and the USA. Volume for volume, neoprene is appreciably cheaper than the high quality insitu sealants such as polysulphides and silicone rubber, but the cost of accurately forming the joint to receive the preformed strip reduces this margin.

When correctly installed, proprietary cellular neoprene strips will remain watertight against pressures of up to 3 atm (30 m head of water). Present indications, based on tests at Northampton sewage treatment works, show that neoprene is the most durable of the sealants and is particularly resistant to attack by bacteria and mould growths.

Insitu Compounds

Insitu compounds are divided into a number of types, namely:

a. Mastics
b. Thermoplastics—hot applied
c. Thermoplastics—cold applied
d. Thermosetting compounds—chemically curing
e. Thermosetting compounds—solvent release

The author is indebted to the American Concrete Institute for much of the information which follows.

a. Mastics. Mastics are generally composed of a viscous liquid with the addition of fillers or fibres. They maintain their shape and stiffness by the formation of a skin on the surface and do not harden throughout the material nor set in the generally accepted use of the term. The vehicles, *i.e.* the viscous liquids, are usually low melting point asphalts, polybutylene, or a combination of these. They are used where the overriding factor is low first cost and where maintenance and replacement costs are not considered important. The extension–compression range is small and so these materials should only be used where small movement is anticipated.

b. Thermoplastics—hot applied. These materials become fluid on heating and on cooling they become an elastic solid, but the changes are physical only and no chemical reaction occurs. A typical example of this type of sealant is the rubber–bitumen compounds which are used extensively in many countries.

As the sealant has to be applied in a semi-liquid state it is only suitable for horizontal joints; it is used largely for roads and airfield pavements, but can also be used in the floors of reservoirs and sewage tanks.

The movement range which this type of material can accommodate is rather greater than that obtained with mastics, but is still small compared with thermosetting chemical curing elastomers.

There is a British Standard for 'Hot Applied Sealing Compounds for Concrete Pavements', BS 2499. The Standard is strictly a performance specification and says nothing about the chemical composition of the sealants. When used for tanks to hold potable water the sealant must be non-toxic and should not contain phenol compounds.

c. Thermoplastics—cold applied. These materials set and harden either by the evaporation of solvents (solvent release) or the breakup of emulsions on exposure to air. Sometimes a certain amount of heat is applied to assist workability, but generally they are used at ambient air temperature.

For water retaining structures the most popular type is a rubber-asphalt. The movement which this type of sealant can accommodate is small; it hardens with age and suffers a corresponding reduction in elasticity. There is no British Standard for this type of sealant. A check should be made to ensure that the sealant does not contain toxic or phenol compounds and will not support bacterial or fungoid growth, before accepting its use in potable water tanks.

d. Thermosetting compounds—chemically curing. Materials in this category are one- or two-component compounds which cure (mature or harden) by chemical reaction to a solid state from the liquid or semi-liquid state in which they are applied.

High grade materials in this class are flexible and resilient and possess good weathering properties; they are also inert to a wide range of chemicals.

These compounds include polysulphides, polyurethanes, silicone rubber and epoxide based materials. They can be obtained so as to have an expansion–compression range of up to $\pm 25\%$ and a temperature range from $-40\,^{\circ}\mathrm{C}$ to $+80\,^{\circ}\mathrm{C}$.

While they are considerably more expensive than the mastics and thermoplastics, they will accommodate far greater movement and are more durable.

In the UK the polysulphides are the most popular and can be used with confidence in a wide range of liquid-retaining structures, including those containing potable water. Some are used with a primer and some without. It is important to ascertain whether the particular brand selected will bond to damp concrete or whether a dry surface is required. Complete adhesion between the sealant and the sides (but not the base) of the sealing groove is essential for a liquid-tight joint. In the climatic conditions of the UK it is virtually impossible to ensure dry concrete on most construction sites. Two-part polysulphide-based sealants are covered by BS 4254, which deals with two grades, a pouring grade and a gun grade. One-part gun grade polysulphide-based sealants are covered by BS 5215.

e. Thermosetting compounds—solvent release. Sealants of this type cure by the release of solvents present in the compound itself. The principal materials in use are based on such compounds as butyl, neoprene and polyethylene. Their general characteristics are somewhat similar to those of solvent-release thermoplastics; their extension–compression range is about $\pm 7\%$.

There is no British Standard for this type of sealant.

1.8 ORGANIC POLYMERS

Organic polymers are complex chemical compounds derived mainly from the petro-chemical industry. These materials are often referred to as 'resins', and the principal resins used in the construction industry are epoxide, polyurethane, polyester, acrylic, polyvinyl acetate, and styrene butadiene. The basic raw materials are supplied by comparatively few manufacturers, such as Shell Chemicals, CIBA, Dunlop Chemical Products Division, Revertex, Borden Chemical Co. and Dow Chemicals. The raw matérials are then taken by a large number of formulators who formulate the final products in such a way that they possess specific characteristics needed for the use to which they are put.

Under the particular condition of curing, which in some cases requires hardeners or accelerators, the resins form long molecular chains in three dimensions, which can result in an extremely strong and stable material.

While the range of use of these materials is now very wide, it is convenient and practical to divide it into two main categories; namely, coatings in which the formulated compound is used on its own, and mortars and concretes in which the resin is mixed with aggregate and sometimes cement.

It is quite usual to find that several types of polymer are used in combination in order to obtain the optimum results. The information given here is intended for practising engineers and not for chemists.

For coatings, used for the protection of concrete and to improve impermeability, epoxies and polyurethanes are in most general use. For use in mortars and concretes, epoxies, polyurethanes, acrylics and styrene butadiene compounds are used successfully. Polyvinyl acetate (pva) is used in large quantities as a bonding agent for floor screeds and toppings to increase adhesion with the base concrete; it is also used in cement mortar mixes to improve certain characteristics of the mortar and bond with the substrate.

1.8.1 Epoxide Resins

The resins are marketed by the formulators and have the special properties required for the specific use to which they will be put. For example, some resins can be successfully applied and cured under water. While most epoxies are rigid when cured, it is now possible to obtain a type which is slightly flexible.

The basic characteristics of epoxide resins include the following:

(a) Outstanding adhesive qualities to such materials as concrete and steel
(b) Resistance to a wide range of acids and alkalis and other chemicals, except acids such as nitric which have high oxidising characteristics
(c) Rather vulnerable to organic solvents
(d) Low shrinkage when the compound cures and changes from the liquid to the solid state
(e) High coefficient of thermal movement compared with concrete
(f) High compressive, tensile and flexural strength
(g) Appreciable loss of strength at temperatures over about 80 °C
(h) High rate of gain of strength, which can be varied to suit the particular application
(i) Rather poor resistance to fire compared with concrete and clay bricks
(j) To obtain satisfactory results, the site conditions under which the resin will be applied must be stated to the formulator. Unless specially formulated, most epoxide resins must be applied to dry surfaces and the ambient air temperature and relative humidity during application must be kept within fairly narrow limits.

The resins are usually two-pack materials, consisting of a basic resin and an accelerator (sometimes called a hardener or activator). The two materials must be thoroughly mixed immediately prior to use.

Some of the more important factors relating to the use of epoxides are:

(1) *Pot life.* This is the period which can be allowed to elapse between the mixing of the resin and the activator and the completion of the application of that particular mix. This period can be varied by the formulator to suit site conditions, but increasing the pot life will result in a slowing down of the rate of hardening of the applied coating. The period usually adopted is fairly short, namely a few hours, but the actual range can be between 30 min and about 48 h.
(2) *Hardening.* This is the physical setting of the plastic resin after application; it can be varied by the formulator. It is generally recommended that each coat must harden before the next coat is applied, and therefore applicators must acquaint themselves with the hardening characteristics of the particular resin they propose to use.

(3) *Curing*. Curing is the term used to describe the maturing and gain of strength of the resin. It is really the formation of the molecular linkage which imparts the desirable strength and durability to the final product. The curing period can also be varied by the formulator, but an 'average' period is about seven days. Curing usually ceases when the ambient air temperature falls to about 5 °C and this can cause some difficulties on site when these resins are used and early strength and bond is required.

There are about fifteen different types of epoxide resins on the market and probably about three hundred hardeners, so that the possible number of combinations of resin and hardener is very high. For this reason it is inadvisable for anyone who is not an experienced polymer chemist to attempt to advise on a specification for an epoxide resin for a particular purpose. The basic requirements for the use of the resin must of course be clearly laid down, such as setting time, curing time, ability to bond to wet concrete etc.; the details of how these requirements are met must be left to the formulator.

1.8.2 Polyurethanes

Polyurethanes, like epoxide resins, are products of the petrochemical industry. They can be obtained as elastomers, solid and rigid materials and as flexible coatings. They are very durable in external conditions and retain their gloss well. For use in the construction industry they are usually formulated as two-pack and single-pack materials. Generally, the two-pack material has better durability than the single-pack. Polyurethanes can be specially formulated to meet specific site requirements. One characteristic of particular importance is that they will cure in temperatures appreciably below 0 °C, whereas epoxide resins cannot be relied upon to continue to gain in strength (cure) when the temperature falls to 5 °C and below. They can be combined with epoxies and will withstand relatively high temperatures as well as sudden changes in temperature, *i.e.* thermal shock.

1.8.3 Polyester Resins

Polyester resins are used with Portland cement and selected aggregates to form a polymer–cement–aggregate mortar. Such mortars possess a number of desirable qualities such as good resistance to a wide range of chemicals, high resistance to abrasion, complete waterproofness under the range of heads likely to be met in liquid-retaining structures, and high bond strength with most common building materials.

They are also used with glass fibre to form linings to various types of liquid-retaining structures and liquid-conveying systems such as large diameter sewers.

There are a number of important differences between the properties of polyester resins and epoxide resins. The polyesters can be used in a wider temperature range; they offer rather better resistance to heat, but have appreciably higher shrinkage characteristics; the bonding properties to concrete are generally lower. By adjusting the ratio of resin to catalyst, the hardening time can be made very short, and once a 'set' occurs the rate of gain of strength is very rapid.

1.8.4 Polyvinyl Acetate (pva)

This material is used as a bonding agent and as an admixture in mortar to improve certain properties of the mortar. Manufacturers of proprietary compounds based on pva claim increased tensile and flexural strength, reduced drying shrinkage and reduced permeability. The author considers that some of the improvements claimed are marginal and the behaviour of pva under permanently damp conditions can be very disappointing.

1.8.5 Styrene Butadiene and Acrylic Resins

These materials are also known as latexes and polymer emulsions. Some of the properties of a proprietary styrene butadiene copolymer emulsion (latex) used with Portland cement in grout, mortar and concrete are:

pH	11·0
Total solid content	47 %
Specific gravity	1·1

The author is indebted to Revertex Ltd for the above information which relates to their latex 'Revenex 29Y40'. Acrylic latexes (styrene acrylic and all acrylic) generally have shorter setting and hardening times and greater resistance to ultra violet light than the styrene butadiene type, but are more expensive.

The advantages claimed for the use of these latexes in Portland cement grout, mortar and concrete, include reduced permeability, reduced initial drying shrinkage, improved resistance to attack by certain dilute acids and solutions of sulphates, and improved bond to the substrate.

Special precautions have to be taken by formulators with styrene butadiene latexes when these are used with OPC to reduce air entrainment and the effect of retardation of the cement. When used in concrete and mortar, a reduction in compressive strength should be expected, but this is unlikely to exceed about 12 % with normal dosage. A further point is that it may be desirable to delay the start of wet curing for 12 to 24 h after placing, and the supplier should be consulted on this matter.

1.9 POLYMERISED CONCRETE

This is sometimes rather loosely referred to as polymer concrete. In this book polymerised concrete means Portland cement concrete containing a monomer and which is polymerised after it has hardened. On the other hand polymer concrete is concrete in which the cement is replaced either entirely or principally by an organic polymer such as epoxide or polyester resin, or normal concrete which contains polymer as an admixture.

Polymerised concrete can be divided into two types:

(1) The complete unit of hardened concrete is impregnated (usually by dipping) with a monomer and is then polymerised either by heat or gamma rays.
(2) The monomer is mixed with the gauging water and then after the concrete unit has hardened it is polymerised by heat.

A considerable amount of work has been carried on for some years in a number of countries into the techniques of polymerising concrete and the characteristics of the finished concrete. Claims made for polymerised concrete include the following:

(a) A considerable increase (up to four times) in the compressive and tensile strength.
(b) The resistance to chemical attack and the effects of freeze-thaw are greatly increased.
(c) Absorption and permeability are greatly reduced.

All the above increases and decreases are related to control specimens of normal Portland cement concrete.

The process of producing polymerised concrete is complicated and expensive and so far it appears to have been used only on a small scale in selected locations for long term test purposes. Some references on this material are given in the Bibliography at the end of this chapter; the Technical Report No. 9 by the Concrete Society contains an extensive bibliography.

1.10 BUTYL RUBBER SHEETING

Butyl rubber is a tough, black flexible sheeting with considerable abrasion resistance. It is described as a copolymer of polyisobutylene together with a small percentage of isoprene. The sheets are used as a waterproof membrane in roof construction. Although it is tough compared with many other flexible sheeting materials, it requires careful laying as it can be readily punctured by sharp tools and projections on the concrete surface. The sheets can be joined together by hot vulcanising, which results in a very strong joint, or by surface adhesives, which give a weaker less durable joint.

The thickness of the sheets varies from about 0·50 mm to 2·00 mm. The black colour of the material is sometimes a disadvantage and considerable efforts have been made to pigment it and to develop a durable method of painting. So far, neither of these has proved entirely satisfactory. The latest development work is directed towards a laminated sheet, using a material which can be readily pigmented as the surface layer.

1.11 POLYISOBUTYLENE

PIB is somewhat similar to butyl rubber, but it has a number of important qualities not possessed by the latter material It is marketed as a black flexible sheeting. It originated in Germany and was first sold in the UK under the trade names of Opanol and Rhepanol. Two of its most important features are its low restitution (it does not harden or crack) and experience over a long period of years has shown that it is not adversely affected by ultra-violet light, ozone, a wide range of chemicals, and does not support fungi growth. The low restitution means that when the material is stretched there is less tendency to regain its original dimensions and thus stress is reduced.

The sheets are normally joined by solvent welding. The solvent softens and activates the material and completely evaporates resulting in a genuine weld having the same strength and durability as the original material. This simple method of jointing enables it to be cut and dressed to fit into corners and around obstructions.

The abrasion resistance of PIB is lower than that of butyl rubber and therefore special precautions must be taken when it is used on surfaces likely to be exposed to foot traffic. The sheets are made in thicknesses of 0·50–2·00 mm. For general purposes, a thickness of 0·8 or 1·0 mm is adequate. As previously mentioned its natural colour is black, and like butyl rubber, it is not readily pigmented or painted.

1.12 GLASS FIBRE REINFORCED PLASTICS

This material is usually referred to as GRP. It is a composite material composed of polyester resin and glass fibre. There are various types of glass fibre and various types of polyester resin. The manufacturers select the most suitable type for a particular application. The resin and the glass fibre are the essential materials and form the bulk of the final product but there are in addition fillers, pigments, stabilisers and numerous other additives to impart special properties to finished material.

The author has not seen any published information on the use of this material for the repair of concrete, but he has seen examples of GRP linings applied to liquid retaining structures to remedy leakage and to protect the concrete against chemical attack.

The process of application consists in building up successive layers of the resin and glass fibre on the substrate. This can be done by hand or spray, but the latter is now more usual. With spray application, the resin, catalyst and glass fibre are applied through a three-nozzle gun. It is important that the glass fibres are completely covered by the resin, and special techniques are employed to ensure this. Normal good practice for the preparation of the concrete surface prior to application of the resin/glass fibre matrix, is essential.

BIBLIOGRAPHY

LEA, F. M. AND DESCH, C. H., *The Chemistry of Cement and Concrete*, 3rd ed., Edward Arnold Ltd. London, 1970, p. 725

SHIRLEY, D. E., Principles and Practice in the Use of High Alumina Cements, *Municipal Engineering*, **145**(4 & 5) (1968)

ANON., No more wetness about warmth, *New Civil Engineer*, 21 March 1974, pp. 40, 43

FRENCH, P. J., MONTGOMERY, R. G. J. AND ROBSON, T. D., High concrete strengths within the hour. *Concrete*, August 1971, pp. 253, 258

THE CONCRETE SOCIETY, Admixtures for Concrete (The Concrete and Reinforced Concrete Committee), *Concrete*, January 1968, 39–46

AMERICAN CONCRETE INSTITUTE, Admixtures for Concrete (Committee 212), *Proc. A.C.I.*, **60**(11) (1963) pp. 1481–1526

AMERICAN CONCRETE INSTITUTE, Guide to Joint Sealants to Concrete Structures (Title 67-31), *Journal A.C.I.*, July 1970, pp. 489–536

INSTITUTION OF STRUCTURAL ENGINEERS, *Guidelines for the Appraisal of Structural Components in High Alumina Cement Concrete.* I.S.E., Oct. 1974

RAGSDALE, L. A. AND RAYNHAM, E. A., *Building Materials Technology*, 2nd ed., Edward Arnold Ltd. London, 1972, p. 336

GEORGE, C. M., *The Structural Use of High Alumina Cement Concrete*, Lafarge Fondu International, France, 1975, p. 15

MIDGLEY, H. G. AND MIDGLEY, A., The conversion of high alumina cement. *Mag. Conc. Res.*, **27**(91), June 1975, pp. 59–77

THE CONCRETE SOCIETY, *Polymer Concrete*, Technical Report No. 9, 1975, Jan., Concrete Society, London, p. 47

THE CEMENT & CONCRETE ASSOCIATION AND THE CEMENT ADMIXTURES ASSOCIATION, *Superplasticizing admixtures for concrete*, C.C.A., London, 1976

ROBERTS, M. H. AND JAFFREY, S. A. M. T., *Rapid chemical test for the detection of high alumina cement concrete.* Building Research Establishment, IS 15/74, HMSO, London, p. 3

LOWREY, K. W., Protective coatings for structural systems, *Civil Engineering* (October, 1972), London, p. 7

SEALANT MANUFACTURERS CONFERENCE, *Manual of Good Practice in Sealant Application*, Sealant Manufacturers' Conference and Construction Industry Research and Information Association, London, Jan. 1976, p. 72

CEMBUREAU, *Cement Standards of the World, Portland Cement and its Derivatives*, Cembureau, Paris, 1968, p. 356

CHAPTER 2

Factors Controlling the Deterioration of Concrete, Steel and Other Metals Used in Concrete Structures

It is intended in this chapter to consider the more important factors which can cause deterioration in reinforced concrete structures. It is only by carefully reviewing why and how this deterioration occurs that satisfactory techniques can be developed for the repair. This review will also be useful in helping to prevent this deterioration from taking place in future structures. The accumulation of knowledge and experience is a continuous process. Mistakes are inevitable; they should also be forgivable provided the necessary lessons are learned.

At the present time there is no material known which is completely inert to chemical action and immune to physical deterioration. While concrete is no exception to this, it has, under what may be termed 'normal conditions of exposure' a very long life. Concrete made from naturally occurring cements (pozzolanic materials such as trass) has been found in excellent condition after more than 2000 years. There is no reason to believe that under similar conditions, modern Portland cement concrete will have a shorter life. This should satisfy even the most conservative client! However it is known that environmental conditions can be aggressive to concrete and other materials used in modern construction.

Only two basic materials will be considered in this Section, namely the metals and the concrete used in reinforced concrete structures. The metals most commonly used are various types of steel (ferrous metals), copper, phosphor bronze, gunmetal, zinc and aluminium.

For the purpose of this Section, the term 'aggressive environment' is intended to mean the environmental conditions under which the structure will operate which may cause a deterioration in the materials of which it is built, and as a consequence, may reduce its useful life. Errors in design, overloading and foundation failures are not included.

2.1 DURABILITY AND IMPERMEABILITY OF CONCRETE

A durable concrete containing steel reinforcement must be impermeable. This is particularly important if conditions of exposure are moderate, severe, or very severe. These terms are intended to have the same meaning as in Code of Practice (CP) 110: The Structural Use of Concrete (Table 19, p. 54). The author has often met divergent views on the practical meaning of the words 'durability' and 'impermeability', and therefore a short discussion on this subject follows.

2.1.1 DURABILITY

A structure would be considered durable if it fulfilled its intended duty for the whole of its design life with the minimum of maintenance. The design life of the structure will usually be laid down by the client in consultation with the designer. It would be unrealistic to expect any structure to maintain its 'as new' condition for a period of 60–100 years without any maintenance whatever.

However, Portland cement concrete has the potential of an almost unlimited life unless it is subjected to chemical attack by an aggressive environment, or suffers physical damage. Weather staining and similar discoloration should not be confused with lack of durability. On the other hand, deep carbonation, chemical attack on the concrete, cracking and spalling due to poor quality materials or workmanship, and/or corrosion of the reinforcement, would be a clear case of low durability.

Concrete structures never consist entirely of concrete, as all these structures contain other materials which vary with the type of structure and include reinforcement, joint sealants and fillers, thermal and sound insulation, metal fixings, pipe connections, and waterproofing and decorative layers. Some of these 'other materials' will have a limited life compared with the concrete. In particular, joint sealants, fillers, waterproofing and decorative layers will require periodic renewal.

The question of how to achieve durability in concrete is discussed in the next Section, in which it will be seen that durability is closely related to impermeability. In some cases, certain parts of a structure may be subject to physical deterioration such as abrasion caused by steel wheeled trolleys on a floor, a jet or stream of high velocity water containing grit impinging on a concrete wall or floor, spalling and surface flaking due to freeze-thaw cycles, and damage by wave action abrasion by sand and shingle in the case of marine structures.

2.1.2 IMPERMEABILITY

The standard of impermeability required for a reinforced concrete structure will be governed by the degree of exposure and whether environmental conditions are aggressive.

A basic consideration in dealing with the problem of impermeability in concrete is to realise that concrete possesses a pore structure, and in this respect is fundamentally different to metals. The capillary pore structure allows water under pressure to pass slowly through the material. The permeability rate through good quality dense concrete is extremely slow. Permeability tests are very difficult to carry out, and laboratory work is confined to low pressures of about 1·0 to 1·5 m head (a tenth to a sixth of an atmosphere).

The factors governing the permeability of concrete are very complicated and even today there are differences of opinion among concrete technologists on the relative importance of the many variables involved. Readers who wish to go into this subject in detail should refer to the Bibliography at the end of the chapter.

However, most concrete engineers agree that the impermeability of concrete depends on many factors, including the following:

(a) The quality of the solid constituents, cement and aggregates
(b) The quality and quantity of the cement paste, which in turn depends on the quantity of cement in the mix, the w/c ratio, and the degree of hydration of the cement
(c) The bond developed between the paste and the aggregate
(d) The degree of compaction of the concrete
(e) The presence or absence of cracking due to primary or secondary stresses
(f) The adequacy of curing

Of these six factors, the w/c ratio probably exercises the greatest single influence on permeability.

The specification should lay down the requirements to obtain maximum impermeability of the concrete. There are differences of opinion among practising engineers as to what such a specification should contain. The two main schools of thought are those who claim that the main clause should be a minimum cube strength, and those who recommend a minimum cement content and maximum w/c ratio, as the dominating requirements.

The author recommends a minimum cement content, a maximum w/c ratio, and a minimum cube strength which should as far as possible be compatible with the cement content and the w/c ratio and type of aggregate likely to be used. The reason for this is:

(a) The fundamental importance of ensuring an adequate quantity of good quality cement paste
(b) The relative ease of cube testing and the fact that this is a well known and accepted method of test.

Until recently, it was very difficult to ascertain the cement content in hardened concrete, but this problem has now been satisfactorily solved by the Cement and Concrete Association's patented equipment for the rapid analysis of fresh concrete. This method is described briefly in Appendix 5.

Regarding cube testing, it is essential to realise that this test only provides information on the quality of the concrete as mixed; it does not, and cannot give direct and accurate information on the quality of the hardened concrete in the structure. This is discussed in more detail in Appendix 4, under Core Testing.

2.1.3 CARBONATION

Carbonation is the effect of carbon dioxide (CO_2) in the air on Portland cement products, mainly calcium hydroxide ($Ca(OH)_2$) in the presence of moisture. The $Ca(OH)_2$ is converted to calcium carbonate ($CaCO_3$) by absorption of carbon dioxide. The calcium carbonate is only slightly soluble in water and therefore, when it is formed it tends to seal the surface pores of the concrete, provided the concrete is dense and impermeable.

The pH of the pore water in concrete is generally between 10·5 and 11·5, but if, due to carbonation it is lowered to 9·0 and below, corrosion of the reinforcement may occur. Therefore the depth of carbonation in reinforced concrete is an important factor in the protection of the reinforcement; the deeper the carbonation, the greater the risk of corrosion of the steel. The extent of carbonation can be determined by treatment with phenolphthalein; the presence of alkalinity shows as a pink colour, while the carbonated part of the concrete remains without colour change.

Good quality dense concrete carbonates very slowly; even after a period of 50 years carbonation is unlikely to penetrate to a greater depth than 5–10 mm. On the other hand, a low strength, permeable concrete may carbonate to a depth of 25 mm in less than 10 years. Experience suggests that low quality cast stone products are particularly prone to carbonation; this is discussed in more detail in Chapter 3.

2.2 CHEMICAL AGGRESSION TO CONCRETE

Chemical attack on concrete is likely to arise from one of the following causes:

(1) Aggressive compounds in the sub-soil and/or ground water
(2) Aggressive chemicals in the air surrounding the structure
(3) Aggressive liquid or chemicals stored in or carried by the structure
(4) The formation of aggressive compounds by bacterial action in direct contact with the concrete

The protection of concrete structures of various types against chemical attack is dealt with in Chapter 6, therefore only brief comments will be given here.

It is fortunate that most soils and ground waters in the UK are not aggressive to Portland cement concrete. When aggression does occur it is usually caused by sulphates in solution in clay soils; apart from sulphates, some naturally occurring ground waters are acidic due to organic acids from peaty soils. Generally these are only mildly aggressive to concrete. However, in industrial tips, the subsoil and ground water can be very aggressive and such areas are potentially dangerous to concrete.

Cases of aerial aggression to concrete are rare, and this type of attack is only likely to occur in chemical factories where the concrete members are located over or adjacent to vats holding liquids which give off aggressive fumes. Most cases of chemical aggression to concrete occur when the structure is used to hold or carry liquids which will attack the Portland cement matrix. Chemicals in the dry state are unlikely to attack concrete. The case of aggressive compounds (usually sulphuric acid) formed by bacterial action is generally confined to sewers and sewage treatment tanks.

2.3 CHEMICAL AGGRESSION TO METALS

An important basic principle when using metals in the repair of structures is that dissimilar metals should not be allowed to be in direct contact. Under the conditions which exist in most structures, electro-chemical action will take place between the two metals if they are in contact. One metal will form the anode and the other the cathode; the anode corrodes, the cathode does not. This is in fact the principle of cathodic protection which is described in Chapter 5. Magnesium, zinc and aluminium are anodic to iron and steel, but iron and steel are anodic to brass, copper, phosphor-bronze and gunmetal. In many instances mild steel will be anodic to stainless steel.

2.3.1 Mild and High Tensile Steel

The metal most used in concrete structures is steel (mild and high tensile) in the form of reinforcement. Steel reinforcement is protected from corrosion by the surrounding concrete provided this concrete is good quality and of adequate thickness. Portland cement concrete provides an intensely alkaline environment around the steel. The pH of the cement matrix is in the range of 10·5–11·5. Research has shown that within this pH range steel is passivated and this inhibits corrosion.

The corrosion of steel is a complicated electro-chemical process. Unprotected steel corrodes in the presence of moisture due to differences in electrical potential on the surface of the steel which form anodic and cathodic areas. The metal oxidises at the anode and corrosion occurs there.

Treadaway and Russell in Building Research Station Current Paper

CP/82/68, give the following three factors which control the corrosion rate of steel reinforcement in concrete:

1. Contact between the steel and the ionically conducting aqueous phase in concrete. This is dependent upon the moisture content and the constituents of the concrete.
2. The presence of anodic and cathodic sites on the metal in contact with the electrolyte. This is a function of metal surface variations (oxide coating in contact with bare metal surfaces) and the environment.
3. The availability of oxygen to enable the cathodic reactions to proceed (cathodic depolarisation).

Cathodic depolarisation is controlled by the rate of diffusion of oxygen from the atmosphere through the concrete matrix to the cathodic sites on the steel surface. From this it can be seen that the extent or rate of corrosion is dependent on the presence of oxygen and the permeability of the concrete. This applies equally to the reduction of the pH of the concrete by the reaction of carbon dioxide with the alkalis in the cement paste. If the concrete carbonates to a depth equal to the concrete cover to the reinforcement, the passivity of the steel will be destroyed and corrosion accelerated. It has been estimated that if the pH falls to 9·0 corrosion is likely to start.

Passivity can also be destroyed by the presence of chloride ions and thus the presence of chlorides in a concrete mix is a potential cause of corrosion. The corrosion potential of chlorides is increased by porous concrete, which in turn can result from inadequate compaction, high w/c ratio, low cement content, poor mix design or inadequate mixing. The natural passivity of the concrete surrounding the reinforcement can in theory be increased, and therefore its protection improved, by the use of corrosion inhibitors.

Corrosion inhibitors can be incorporated in the concrete, but in fact this course is not recommended for a number of reasons which are discussed in the Current Paper mentioned above. A better method, but more difficult to arrange on site, is to include the selected inhibitors in a cement grout which is applied as a coating to the reinforcement. The work done by Treadaway and Russell suggests that a combination of sodium nitrite and sodium benzoate is likely to give the best results using readily available and fairly inexpensive chemicals. The author's experience is that cement grout applied to reinforcement tends to dry rather quickly and flake off but this can be substantially improved by the substitution of a styrene butadiene latex for the gauging water. Even so, it is important that the concrete or mortar should be placed as soon as practical after the coating of grout has hardened.

An alternative, more durable, but more expensive coating is to use epoxide resin; however, the additional expense is only justified in extreme cases. This is discussed in some detail in Chapter 5.

2.3.2 Removal of Rust from Reinforcement

Mill-scale is formed under mildly reducing conditions during the manufacture of the bars and consists of oxides of iron, with the lower oxides, FeO and Fe_2O_3, predominating, but the higher oxide, Fe_3O_4 is also likely to be present. Mill-scale is defined in Clause 102 of CP 2008, as 'stratified layers of the oxides of iron, FeO, Fe_2O_3, and Fe_3O_4.' As the mill-scale weathers, it combines with oxygen in the air, *i.e.* it oxidises and increases in volume and tends to lose its adhesion to the steel below. Generally mill-scale is not continuous over the whole bar but is present as patches. It is this discontinuity which assists the formation of corrosion cells due to the difference in electrical potential between the mill-scale and the steel.

The question then arises as to what extent it is necessary to remove mill-scale from reinforcement. There are wide divergences of view on this matter, and as with many other problems, it is not advisable to be too dogmatic, but to use common sense. To be more specific, the author recommends that loose mill-scale should be removed, but there is no need to remove scale which is adhering firmly to the bars.

Similar principles can be applied to rust on reinforcement. Light powdered rust which does not come off when the bars are hit, can be left. Some engineers claim that this type of rust will improve bond between the steel and the concrete or mortar. However, for the reasons given above, the presence of rust may slightly increase the risk of corrosion if the passivity of the steel is reduced due to ingress of moisture, carbonation, presence of chlorides etc. All rust which is in the form of scale or flake should be removed before new concrete or mortar is placed.

A popular practice with some contractors who carry out repairs to concrete, is to use rust inhibitors, based on phosphoric acid, on the steel reinforcement after the rust has been removed. The author's experience is that the use of such inhibitors can encourage the workmen to take less care with the removal of the rust. The phosphoric acid reacts with the steel to form iron phosphate which helps to protect the metal against corrosion. However these acid inhibitors are not recommended for the following reasons:

(a) If the rust is properly removed they are not necessary as the new concrete or mortar will passivate the steel.

(b) Because they are acid and the surrounding concrete is alkaline they will attack any concrete with which they come in contact and will neutralise its alkalinity. Therefore a special 'anti-acid' coating has to be applied over the inhibitor, which means another activity for the workmen to carry out and which should be supervised.

(c) If the acid inhibitor is applied too generously it will run down the reinforcement and penetrate into the sound concrete in which the steel is embedded.

It is better to concentrate on the proper removal of the rust and any loose mill-scale, and then apply a good coat of grout containing the non-aggressive inhibitors already mentioned to the steel, or to use a grout made with Portland cement and a styrene butadiene latex emulsion (2 parts cement to 1 part emulsion by weight).

2.3.3 Galvanised Reinforcement

Moist Portland cement concrete, because of its alkalinity, will attack zinc and aluminium. It is therefore advisable to protect with bituminous paint or similar coating, any aluminium or zinc coated steel which is in continuous contact with damp concrete.

When galvanised reinforcement is used in concrete, there will be some initial attack on the galvanising, but this is unlikely to be significant in good quality impermeable concrete. Some Portland cements contain chromates; if the chromate exceeds about 65 ppm, this will inhibit the reaction between the cement paste and the zinc by the formation of a passive layer of zinc chromate on the surface of the galvanising. If chromates are not present in the cement, then it is relatively simple to add them in the galvanising vats. In repair work, the use of galvanised reinforcement is sometimes justified, particularly when external conditions are aggressive, degree of exposure is high, or the cover less than that normally required. The advantage of galvanised reinforcement, when cover to reinforcement is reduced, is recognised in BS 1217: 1975: Cast Stone.

2.3.4 Stainless Steel

Stainless steel has been discussed briefly in Chapter 1; it is important to select the right grade for the anticipated conditions of exposure. For severe conditions it is advisable to specify type En58J (also known as type 316 steel), which is an austenitic steel; the normal use is for dowel bars and fixings as these are often important features of concrete repair work. It is prudent to take expert advice, from such an organisation as the Corrosion Advice Bureau of the British Steel Corporation, when it is necessary to have two metals of different chemical composition in contact, as this can give rise to electrolytic action resulting in the corrosion of the metal which forms the anode.

2.3.5 Aluminium

If unanodised aluminium is used in contact with damp concrete it should be protected by a thick coating of bituminous paint or other material which will not be attacked by the caustic alkalis in the cement. The durability, resistance to corrosion, and appearance of aluminium is improved by anodising. A brief note on anodising is included in Chapter 1.

Generally, aluminium is anodic to steel, but this does not necessarily apply to some alloys of aluminium as these may be cathodic to steel. Care must therefore be exercised if steel and aluminium are in direct contact.

2.3.6 Copper

Copper is resistant to attack and is very durable in the conditions met in most building structures. It is not corroded by Portland cement concrete unless chlorides are present. Copper is generally cathodic to steel and therefore it is important that these two metals should not be in direct contact, otherwise corrosion of the steel is likely to occur.

2.3.7 Phosphor-Bronze and Gunmetal

Bronze is a copper–tin alloy, and phosphor-bronze usually contains phosphorus as copper phosphide. Gunmetal is bronze which contains about 9% tin.

These alloys are used for fixings in structures containing aggressive liquids or where regular maintenance is not practical.

2.4 PHYSICAL AGGRESSION TO CONCRETE STRUCTURES

Physical aggression (wear and damage) to concrete can arise from a number of causes, the principal ones being:

1. Freezing and thawing on the outside of structures located in very exposed positions in the Northern part of the British Isles and other countries with similar or more severe climate
2. Thermal shock caused by a sudden and severe drop in the temperature of the concrete, such as spillage of liquefied gases
3. Abrasion to concrete, such as that caused to floors in industrial buildings by steel wheeled trolleys. Similar damage can be caused to the inside of silos, bins and hoppers containing coarse granular material
4. Damage from high velocity water. This damage can be subdivided into three types: cavitation; abrasion from water containing grit; impact from a high velocity jet
5. Abrasion in marine structures caused by sand and shingle thrown against the structure by heavy seas and gale force winds. This is dealt with in Chapter 5.

2.4.1 Freezing and Thawing

Concrete structures which are located in very exposed positions in the North of the British Isles and other countries with a severe climate, may suffer damage by surface scaling of the concrete due to frost action. This disruption of the concrete surface is caused by the penetration of the surface layers by moisture when this is followed by sub-zero temperatures. The consequent freezing and expansion of the water absorbed in the concrete can cause disintegration of the surface layers.

Such conditions with resultant damage are fortunately rare in the UK. It should be noted that similar trouble does occur to marine structures in extreme Northern and Southern latitudes.

Suitable repair measures are difficult to devise, but concrete and mortar used for repairing such damage should be air-entrained. Some basic information on air-entraining admixtures is given in Chapter 1.

Hydraulically pressed precast concrete products have proved their ability to resist the disintegrating effects of freeze-thaw, but it would only be in a limited number of special cases that they could be used for the repair of a concrete structure.

2.4.2 Thermal Shock

Damage to concrete by thermal shock is rare and occurs only in a few special cases such as the spillage onto the concrete of liquefied gases. The gases involved are those used in various industrial processes and include methane, oxygen, nitrogen and hydrogen. The temperatures of these liquefied gases are approximately, −160 °C, −183 °C, −196 °C and −253 °C respectively. Generally the damage is confined to floors but in many cases these floors are subjected to moving loads and therefore the effect of the damage can be more serious.

The author knows of no method of repair to the concrete which can be relied upon to give a long life, and therefore the need for fairly frequent repair must be accepted. The use of high quality air-entrained concrete made with either a limestone aggregate or an artificial lightweight aggregate is likely to give the best results.

2.4.3 Abrasion of Concrete

The abrasion of concrete such as occurs on industrial floors and in silos, bins and hoppers containing certain types of granular material, is a subject on which there is an unfortunate lack of precise information and knowledge. It has been established that the quality of the concrete in terms of compressive strength is the most important factor in determining the abrasion resistance of the concrete surface. The type of finish imparted to the surface is also important, but much less is known about the relative influence of different methods of surface finishing on abrasion. Most problems arise with industrial floors. The use to which the floor is put varies greatly from one industry to another.

After completion of compaction of a floor slab, the following types of direct finish are in general use:

1. No further finish after final pass with the vibrating tamper; this results in a slightly ribbed finish
2. Skip or bull float
3. Power float
4. Hand float (hand trowelling)
5. Early surface grinding following 2 or 3 above

Assuming that all the methods are executed properly, and with a standard high quality concrete, the best wearing and abrasion resistant surface is likely to result from hand trowelling with a steel trowel.

The aggregate does have some influence on abrasion resistance but is only significant if the aggregate quality is particularly poor or especially good. The use of certain types of limestone can result in polishing and slipperiness.

Detailed information on methods of repairs to concrete floors is given in Chapter 3.

2.4.4 Damage to Concrete by High Velocity Water

Damage by high velocity water can be divided into three main categories: cavitation; abrasion from water containing grit etc.; impact from a high velocity jet.

Cavitation

The factors involved in abrasion and impact are fairly well understood but the exact causes of cavitation have not been completely resolved in spite of the research which has taken place in various countries.

A simple, and consequently incomplete explanation of cavitation follows. With high velocity flow, *i.e.* exceeding about 15 m/s, depressions and irregularities in the boundary surface can cause cavitation eddies to form. This is particularly liable to occur down-stream of boundary discontinuities. Fig. 2.1 shows cavitating eddies down-stream of a cylinder located in the cavitation test rig at Imperial College, London. It should be noted that no surface is absolutely smooth, and with very high velocities and under certain special conditions, severe cavitation damage can occur even to steel and other metals. If the absolute pressure at points of surface irregularities approaches the vapour pressure of the water, minute bubbles will form and quickly collapse. The collapse of the bubble wall can produce minute water jets having extremely high velocities. The result of the collapse of the bubbles can create an intense impact wave in the form of a series of hammer-like blows in extremely rapid succession. The effect is very destructive to high strength concrete, and as stated above, even to metals. Readers who wish for a more detailed explanation and discussion on cavitation should refer to the Bibliography at the end of this chapter.

It is obvious that if the formation of the cavitation eddies and bubbles can be prevented, then cavitation will not occur. Generally, damage by

FIG. 2.1. Equipment used for research on cavitation showing cavitation eddies forming and collapsing (courtesy: Imperial College, London (M. J. Kenn)).

cavitation occurs on the surface of spillways, penstocks, aprons, energy dissipating basins, and syphons and tunnels carrying high velocity water.

Cavitation damage to concrete can be fairly easily distinguished from normal erosion, as the damaged areas have a jagged appearance, while erosion from grit laden water results in a comparatively smooth surface. Figure 2.2 shows typical damage to concrete by cavitation in a test rig at Imperial College. In practice, in structures such as spillways, cavitation damage can take the form of the tearing out of large pieces of concrete.

FIG. 2.2. Damage to concrete by cavitation in test rig (courtesy: Imperial College, London (M. J. Kenn)).

Abrasion from Water Containing Grit

High quality concrete is very resistant to damage by abrasion from fast flowing water containing grit. The factors which decide whether any significant damage will occur and the degree of damage, depend on a number of factors of which the following are the more important:

1. The quality of the concrete in terms of compressive strength, cement content and wear resistance of the fine and coarse aggregate
2. The velocity of the flowing water
3. The quantity of the grit carried and the abrasive characteristics of the grit

4. The flow characteristics, that is, whether the flow is continuous or intermittent and the extent to which the quantity of grit varies, hour to hour, day to day etc.

As far as the author is aware there is no published information which shows in practical terms how the above four factors are interrelated. In the circumstances, designers have to use their engineering judgement when deciding whether special precautions are justified in the first instance and if so what these should be. This also applies when damage has occurred and repairs have to be carried out. Obviously the detailed method of repair and the techniques used will depend on the type of structure which has been damaged, the extent of the damage, and the period over which the erosion has taken place.

Impact from a High Velocity Jet

The effect of a very high velocity jet of water striking a concrete surface is to erode the cement paste, resulting in the loosening of the fine and coarse aggregate, and this can lead to the boring of a hole through the concrete. This is the principle of cutting concrete with high velocity water jets which is described in Chapter 5. The pressure at the nozzle immediately before discharge can vary from 400 to 800 atm (6000–12 000 lb/in^2). In fact high quality concrete has proved to be very resistant to erosion by clear water, but it has not been possible so far to fix any limiting factors with regard to critical velocities and concrete strengths. When erosion by a jet has occurred, probably the best solution is the provision of a steel plate securely fixed to the concrete wall or floor. The fixing detail needs careful thought and the use of epoxide mortar is generally desirable.

In cases where the damage is slight and has taken place over a fair period of time, normal concrete repair techniques can be used which are described later in this book.

2.5 THE USE OF SEA WATER FOR MIXING PORTLAND CEMENT CONCRETE AND MORTAR

In most parts of the world there is no need to use sea water for mixing concrete and mortar. However there are some areas where fresh water is so scarce that sea water has to be used if construction work is to be carried out at all. Under these circumstances its use should be restricted if at all possible to plain concrete, but unfortunately most constructional use of concrete requires the addition of steel reinforcement. Sometimes brackish water is available, the salinity of which may be much lower than sea water.

It has been found that in many of the areas where sea and/or brackish water must be used for mixing, the aggregates, particularly the sand, are also contaminated with salt. What then are the dangers of using sea water for mixing concrete and mortar?

In the first place, HAC should not be used for concrete and mortar if the gauging water is highly saline. The manufacturers of this cement should be consulted in all cases where normal fresh (drinking quality) water is not available for mixing.

The following discussion refers to Portland cement concrete and mortar. The total dissolved solids (salts) in Atlantic water is about 32 000 ppm (mg/litre). Of these about 2000 ppm are sulphates (mostly magnesium sulphate), and 18 000 ppm are chlorides (principally sodium chloride). It has been established that the sulphates in normal Atlantic water used for mixing Portland cement concrete and mortar do not result in any significant reduction in the strength and impermeability of the concrete in the long term. The effect of the chlorides is to accelerate the setting and hardening of the cement.

The overall effect of the 32 000 ppm of salts in the mixing water would be to produce considerable efflorescence on the exposed surfaces of the concrete members. Sodium chloride is very aggressive to ferrous metals, and therefore a detailed consideration of the effect of the chloride concentration may be of interest.

As previously stated, the chloride content of Atlantic water is about 18 000 ppm; in the Persian Gulf, Red Sea etc. the concentration is likely to be about 22 000 ppm or 2·2% by weight. If the w/c ratio of the mix is 0·45, then the chloride content of the mix expressed as a percentage of the weight of the cement would be:

$$0{\cdot}45 \times 2{\cdot}2 = 0{\cdot}99\%, \text{ say } 1\%$$

This is the chloride content and to convert this to the more generally used calcium chloride content, the calculation is as follows:

$$\text{Calcium chloride} = CaCl_2 = 40 + 2(35{\cdot}5) = 110$$

If the maximum permitted dosage of anhydrous calcium chloride in concrete, under the most favourable site control conditions is 1·5% by weight of the cement, the equivalent permitted dosage of chloride would be:

$$\frac{70}{110} \times 1{\cdot}5 = 1{\cdot}0\%$$

It has been noted that in those areas where it may be necessary to use sea water for mixing, the aggregates, particularly the sand may also contain an appreciable quantity of salt. From this it can be seen that there is a danger of corrosion of the reinforcement if sea water is used for mixing the concrete.

A further matter for consideration is that in a marine environment it is likely that the salt content of the concrete will increase in the course of time due to penetration by wind driven spray.

An article in Materials Protection (November 1965) by D. F. Griffin, quotes experimental work on the use of saline mixing water and a wind driven spray environment. Unless the structure has been specifically designed for temporary use and only a relatively short life is expected, then consideration should be given to methods of eliminating or reducing the chloride attack on the reinforcement. The measures to be taken will depend on practical possibilities but should include as many of the following as possible:

(a) Provide a dense, impermeable concrete with adequate cover to the steel. In specification terms this means a minimum cement content of 360 kg/m^3 of concrete, a maximum w/c ratio of 0·45, a minimum cover of 50 mm, and careful, thorough compaction and curing of the concrete. The author considers that the maximum cover should not generally exceed 60 mm, otherwise there may be too great a thickness of concrete without crack control steel.
(b) Special care should be taken in the design to limit crack widths on exposed faces to 0·10 mm.
(c) Wherever possible, it is recommended that galvanised reinforcement should be used, and in this case, Portland cement with a minimum chromate content of 65 ppm should be obtained if possible. If such a cement is not available, then chromates can be added to the galvanising vats.
(d) As an alternative to (c), two good coats of epoxide resin can be applied to the reinforcement, after the removal of rust and loose mill-scale. The hardened epoxide coating will reduce bond, but this can be either allowed for in the design (if bond is critical), or high bond steel can be used.

In concluding this short discussion on the use of sea water for mixing concrete, the work by J. D. Dewer, of the Cement and Concrete Association must be mentioned. Dewer found that the compressive strength of the concrete was improved by the use of sea water at all ages from one day to four years, compared with the same mix made with tap water. It was also found that the sea water concrete was less affected by drier and shorter curing conditions.

It should be noted that although HAC concrete, when correctly proportioned and cured, is very durable in sea water, on no account should sea water be used for mixing concrete or mortar made with this cement.

Reference should be made to Chapter 5 which deals with the repair of marine structures.

BIBLIOGRAPHY

TERZAGHI, H. AND PECK, R. B., *Soil Mechanics in Engineering Practice*, 2nd ed., John Wiley & Sons, New York; Chapman and Hall, London, 1968, p. 752

KALOUSEK, G. L., PORTER, L. C. AND BENTON, E. J., *Cem. Conc. Res.* (2) 1972, pp. 79–89

HALSTEAD, P. E., *Behaviour of structures subjected to aggressive waters.* Paper at Inter-Association Colloquium on Behaviour in Service of Concrete Structures, Liege, June 1975, p. 11

BUILDING RESEARCH ESTABLISHMENT, *Concrete in Sulphate-Bearing Soils and Ground-Waters.* Digest, 174, 1975, HMSO, London, p. 4

AMERICAN CONCRETE INSTITUTE, Guide for the protection of concrete against chemical attack by means of coatings and other corrosion-resistant materials, Report by ACI Committee 515, *Journal A.C.I.* Dec. 1966, pp. 1305–1390

LEA, F. M. AND DESCH, C. H., *The Chemistry of Cement and Concrete*; 2nd edn. (Revised), Edward Arnold Ltd, London, 1970, p. 637

AMERICAN CONCRETE INSTITUTE, *Manual of Concrete Practice.* Part 3—Products and Processes, 1972, A.C.I., Detroit, pp. 515–23 to 515–73

PERKINS, P. H., *The Use of Portland Cement Concrete in Sulphate Bearing Ground and Ground Water of Low pH*, Cement and Concrete Association, London, ADS/30, Apr. 1976, p. 16

TOMLINSON, M. J., *Foundation Design and Construction*, 2nd ed., Pitman Publishing Ltd, London, 1969, p. 785

SHACKLOCK, B. W., *Concrete Constituents and Mix Proportions*, Cement and Concrete Association, London, 1974, p. 102

MURDOCK, L. J. AND BLACKLEDGE, G. F., *Concrete Materials and Practice*, 4th ed., Edward Arnold Ltd, London, 1968, p. 398

BUILDING RESEARCH ESTABLISHMENT, *The Durability of Reinforced Concrete in Sea Waters*, Twentieth Report of the Sea Action Committee of the Institution of Civil Engineers National Building Studies, Research Paper No. 30, 1960, HMSO, p. 42

HOUSTON, J. T., ATIMTAY, T. AND FERGUSTON, P. M., *Corrosion of Reinforcing Steel Embedded in Structural Concrete.* Research Report 112-1F, Centre for Highway Research, University of Texas at Austin, US. March 1972, p. 131

PERKINS, P. H., *Floors—Construction and Finishes*, Cement and Concrete Association, London, 1973, p. 132

AMERICAN CONCRETE INSTITUTE, *Manual of Concrete Practice.* Part 1—Erosion resistance of concrete in hydraulic structures, 1968, A.C.I., Detroit, pp. 210–1 to 210–13

KENN, M. J., *Factors Influencing the Erosion of Concrete by Cavitation*, CIRIA Technical Note No. 1, London, July 1968, p. 15

VICKERS, A. J., FRANCIS, J. R. D. AND GRANT, A. W., *Erosion of Sewers and Drains*, CIRIA Research Report No. 14, London, October 1968, p. 20

BICZOK, I., *Concrete Corrosion and Concrete Protection*, 4th ed., Hungarian Academy of Sciences, 1967, p. 543

HAUSMAN, B. A., Criterion for cathodic protection of steel in concrete. *Materials Protection*, **8**(10) (1969), pp. 23–25

EVERETT, L. H. AND TREADAWAY, K. W. J., *The Use of Galvanized Steel*

Reinforcement in Building, Building Research Station, Current Paper CP 3/70, January 1970, p. 10

TREADAWAY, K. W. J. AND RUSSELL, A. D., *Inhibition of the Corrosion of Steel in Concrete*, Building Research Station, Current Paper CP 82/68, December 1968, p. 5

BRYSON, J. O. AND MATHEY, R. C., Surface condition effect on bond strength of steel beams embedded in concrete. *Proc. A.C.I.*, **59**(3) (1962), pp. 397–406

COOK, A. M., *A Comparison of the Bond Strength of Black and Galvanized Plain Reinforcing Bars to Concrete.* (A confidential report, not published.) The Cement and Concrete Association of Australia TM95, April 1974, p. 16

CLARK, R. R., Protection of concrete from cavitation damage. *Proc. I.C.E.* (May 1971), Technical Note 48, p. 5; Discussion, *Proc. I.C.E.* Dec. 1971

KENN, M. J., Bonneville dam stilling basin. *Journal A.C.I.*, **52** (April 1956), pp. 821–837

ANON., Polymerized fibrous concrete to fix dam spillway as research continues on other applications. *Engineering News Record*, Jan. 9, 1975, p. 10

HALSTEAD, P. E. *et al.*, Durability of Concrete. Eight articles in the Supplement to the *Consulting Engineer* (April/May 1971)

CHAPTER 3

Repairs to Concrete Building Structures—Part 1

In this book consideration of repair is limited to two types of structure, namely buildings and civil engineering structures. There is no clear dividing line between these two, and therefore the author has had to be arbitrary in the placing of structures in each class.

The number of normal buildings greatly exceeds the number of civil engineering structures, and for this reason alone, it is logical that the amount of repair work would be greater. On the other hand, the environmental and operating conditions of buildings are appreciably less arduous than those of most civil engineering structures.

Failures and defects in concrete buildings can be placed in five main categories:

1. Structural deficiency resulting from such causes as error in design, errors in construction, impact, explosion, and change of use resulting in higher live loading than was originally allowed for in the design
2. Fire damage; this often results in some weakening of the structure as a whole as well as severe physical damage to the individual concrete members (floor slabs, beams, columns etc.)
3. Deterioration due to poor quality concrete, inadequate cover to reinforcement, and the presence of chlorides in the concrete
4. Chemical attack on the concrete and/or reinforcement
5. Physical damage caused by the use to which the building or part of the building was put, such as the abrasion of a floor slab in a factory and the abrasion of silos and hoppers holding coarse granular material

The first step in investigating the deterioration of a building structure is to determine the cause. In serious cases urgent action may be needed to provide temporary supports to guard against possible collapse.

The cause of the trouble may be obvious, such as fire, explosion, impact damage etc. However in the majority of cases a detailed investigation is required to establish the reason for the deterioration. After this, considerable experience is needed to decide on the remedial measures most likely to be effective over a long period and which will be reasonably economic in cost.

This chapter contains a detailed discussion with recommendations for the repair of defects falling under categories 1, 2, 3, and 5. Defects under category 4 are dealt with in Chapter 6.

The need for care in any investigation cannot be over-emphasised as correct diagnosis is essential if a satisfactory technical and economic solution is to be found. An architect or engineer who is asked to carry out an investigation into what is often erroneously termed a 'building failure' may find himself involved in litigation concerning the responsibility for the alleged failure and for the cost of rectifying it. This is of course a very wide subject and all that the author can say is that it is in the real interests of both parties to the dispute that every effort be made to settle it without resort to law. Arbitration, which can and should, be an ideal way of settling disputes quickly and privately, may, unfortunately, be subjected to delays and expense similar to those in the High Court. The reason for this is that the procedure under the Arbitration Act 1950 is closely related to that of the High Court. The Arbitrator has far less authority over the parties during the interlocutory proceedings than the Court; in addition, one or both parties can engage highly paid counsel. Obviously if one party does this the other may be forced to do the same in self defence. The 'case stated' procedure, which is obligatory for arbitrations held in England (but no longer so in Scotland), is another cause of delay and thus increase in cost. This in fact is one of the major criticisms made by persons on the Continent against arbitration procedure in this country.

3.1 GENERAL RECOMMENDATIONS FOR INVESTIGATIONS

The following steps are recommended to be taken when carrying out an investigation of a deteriorated concrete structure:

(a) A careful visual examination of the structure should be carried out. The use to which the building or parts of the building has been put is also important. A good pair of binoculars can be very useful at this stage in checking external conditions of the structure at high level.

(b) It may be necessary to supplement information obtained under (a) by an examination of the architect's and engineer's drawings and specifications. Depending on circumstances, it can be helpful if contact is made with the original main contractor and suppliers of precast concrete units (if the latter were used and show signs of deterioration).

(c) Where there are signs of corrosion of reinforcement, the depth of cover should be checked by an electromagnetic type covermeter. The covermeter survey should extend well beyond the visibly defective areas, because if the cover is inadequate, rusting may have started but not advanced sufficiently to cause spalling of the concrete. The results of the survey should be recorded on a drawing.

(d) Samples of concrete should be taken and examined by an experienced concrete testing laboratory with the object of determining: (i) The general characteristics of the concrete including such items as cement type and content, w/c ratio, type and grading of aggregates, efficiency of mixing; (ii) depth of carbonation; (iii) the detection of chlorides and an estimation of their concentration, expressed as a percentage of the cement; (iv) any evidence of chemical attack on the concrete and identification of the aggressive chemicals. This is dealt with in detail in Chapter 6.

(e) If the analysis in (d) (iii) showed chlorides to be present other than a trace, then it is desirable for additional samples to be taken and tested. Chlorides are never found uniformly distributed in concrete and therefore their presence in concentrations higher than about 0·25% by weight of the cement, should be considered as a danger signal justifying further investigation. This is particularly important when the concrete is prestressed or post-tensioned. The problems associated with the repair of concrete containing dangerously high concentrations of chlorides is discussed in a separate section in this chapter.

(f) In some cases there may be comparatively little spalling but an appreciable amount of cracking. Some of the cracks may show rust stains while others do not. Non-structural cracking, not caused by corrosion of reinforcement and accompanied by spalling, usually occurs in the early life of the structure and may be plastic, thermal contraction, or drying shrinkage.

The above list is intended to indicate the sort of information which is likely to be needed in most cases. In a few investigations the type of cement used may be important and it may be necessary to distinguish between ordinary Portland and sulphate-resisting Portland, and in others it may be required to detect if high alumina cement has been used in the concrete.

The test for sulphate-resisting Portland cement in concrete and mortar is difficult and time consuming and until recently the same applied to HAC. However, a simple chemical test for use on hardened concrete has been developed by the Building Research Station at Garston which will give a positive result if HAC has been used. This test may also give a positive result with some types of Portland cement so that it can be considered as a 'fail safe' test. If on test there is a negative result then it can be assumed that HAC has not been used. Further information on HAC is given in Chapter 1.

After the conclusion of the investigations outlined above and determination of the principal causes of the deterioration, consideration should be given to the type of repairs needed, the materials to be used, and the method of carrying out the work. The work can be executed by the direct labour of the client, by competitive tender based on specification, bills of quantities (or schedule of work or schedule of rates), or a package deal type of contract.

The selection of the best method of carrying out the repairs taking into account all relevant circumstances, is of fundamental importance in ensuring a satisfactory long term job. Because of the importance of making a correct decision, the following points are put forward for consideration.

The direct labour method is only applicable when there is a suitable direct labour force available with previous experience in the repair of concrete structures. It may be competitive in those rare cases where the repairs are so complex that neither a consultant nor a specialist contractor can give any realistic estimate of cost.

For practical reasons, the choice is usually between some form of competitive tender based on contract documents prepared by an engineer, or competition between specialist contractors who put forward their own priced proposals in detail, including any guarantees.

Each of these two methods has advantages and disadvantages and this is discussed below.

Competitive Tenders Based on Contract Documents

The obvious and undoubted advantage of this method is that the contractors can be selected from an 'approved' list and they all tender on the same basis and adjudication is therefore simpler. However, few consultants have the necessary knowledge and experience to prepare a detailed specification for materials and workmanship, and this is a major and practical objection to this method. A further, and important point is that some specialist firms will offer a guarantee of a virtually maintenance-free period up to ten years provided they tender on the basis of their own assessment of the detailed work to be done and are allowed to carry it out to their own specification. They will not give a guarantee if they have to do the work in accordance with a specification prepared by the client or the client's technical adviser.

Design and Construct (Package Deal) Contracts

Provided the firms invited to submit detailed proposals, specifications and prices are selected on the basis of proven records of successful jobs, this method can give very satisfactory results. The client has the benefit of the experience and know-how of firms who have specialised in the field for many years. As mentioned above, guarantees will often be given for satisfactory performance for periods up to ten years, with an option to extend the period subject to certain specified conditions.

Some contractors use their own materials which are specially formulated to meet site conditions. This helps to ensure that there is no division of responsibility between the supplier of the materials and the firm executing the work. While the main contractor may be legally responsible, with this type of specialised work the fewer the loopholes the better. The disadvantage of this method is that each offer is based on a different solution to the problem, with regard to both method of execution and the materials to be used.

It is here that the consultant can play an invaluable part in selecting the most suitable solution. Price is only one of the many variables which have to be considered in deciding on the adjudication of a package deal tender.

It would appear to be very difficult to formulate clear recommendations to a client, but in practice this is not necessarily so. The most important factor in making such recommendations is that the client should get the best value for his money. This seldom coincides with spending the least money for the job.

When considering deterioration and consequent repairs to a concrete structure there are likely to be differences between the 'symptoms' of the trouble in reinforced concrete and prestressed concrete. With reinforced concrete, corrosion of the reinforcement will show itself in the early stages by rust stains and cracking and spalling of the concrete. Under normal conditions this process continues at a slow rate and the rust stains etc. give clear visual warning that deterioration has started and that repairs should be put in hand.

With prestressed concrete however, corrosion of the comparatively small diameter wires and strand is unlikely to result in rust staining, cracking and spalling of the concrete. It is extremely unlikely that all the wires will corrode at the same rate, and therefore the corrosion will result in deflexion of the member. Thus visual warning of the reduction in strength of the member will be given. Undue deflexion of beams and slabs should always be looked upon as a warning, and action taken to determine the cause.

It is normal practice to take maximum care in the design and construction of prestressed concrete members. Considering the very large number of prestressed structures of all types constructed during the past 25 to 30 years, the number of failures has been extremely small.

In 1972 there was a sudden collapse of a school roof in London which had been constructed in HAC concrete. The author considers that the near panic which followed this unfortunate incident was unjustified. It is hoped that the investigations and further research into HAC concrete which resulted from this collapse will lead to the formulation of practical recommendations for the structural use of this cement. In turn, this should result in HAC concrete being again accepted as a structural material in the circumstances where its high cost is justified. In the past this has been largely in the precast industry where rapid turn-round of moulds is required.

3.2 THE CAUSES AND TREATMENT OF CRACKS IN CONCRETE MEMBERS

Deterioration of a concrete structure is invariably accompanied by cracking to a greater or lesser degree. The study of cracks in concrete, their cause, significance and their repair is a most important factor in the overall repair and strengthening of concrete structures.

It is important to realise that cracks as such are not necessarily defects nor indicative of a failure in the normal use of the words. The cause of the crack, its width, position, and degree of exposure, are the factors which will decide whether it is in fact a defect and how serious it is.

The cause of the cracking can only be decided by a careful investigation, but its width, position in the member and the degree of exposure can all be easily determined. A question which follows from the above is whether there is an 'acceptable' crack width. By this is meant a width which, under the given circumstances of exposure, can be considered as insignificant and therefore needs no repair or treatment. There is no precise and generally agreed answer to this.

British Standard Code of Practice, CP 110: The Structural Use of Concrete, recommends a maximum crack width of 0·3 mm for general conditions, but this width is reduced to '0·004 times the nominal cover to the main reinforcement in cases where members are exposed to particularly aggressive environmental conditions'. This latter condition means that with a cover of 40 mm to the main steel, a maximum crack width of 0·16 mm is considered acceptable. Crack widths are measured at the surface of the concrete and it is assumed that they reduce in width fairly rapidly as the crack extends inwards. This assumption is no doubt correct when the cracks are caused by tension in the tension zone of a stressed member, but may not hold for cracks caused by thermal contraction and plastic settlement. It is very difficult to measure the width of a crack with any real degree of accuracy under site conditions.

An important practical question in connection with acceptable crack widths is how long that width will be maintained. Will the crack tend to close up or to gradually widen with the passage of time? When the crack is caused by some temporary over-load it will close again when this excess stress is reduced. However, when the cause is permanent then it is likely that the crack will not close and under external conditions there will be a definite tendency for the width to increase with age.

Beyond a certain width, moisture is likely to penetrate into the crack, even for a short distance, and then if the temperature falls below 0 °C, this moisture will turn into ice and the resulting expansion will tend to cause spalling along the edges of the crack. This will occur every time the freeze-thaw cycle is repeated. The crack will then gradually widen and eventually moisture will reach the reinforcement and corrosion will start. The corrosion products (rust) occupy a larger volume than the original metal and the expansive forces will crack and spall the concrete.

Certain factors may operate to counteract the widening of cracks described above. One is compressive stresses which may develop at right angles to the line of the crack, such as occur in the compression zone of structural members. Another is the leaching of lime from the cement paste caused by the movement of moisture through the concrete and the continued hydration of cement particles. This phenomenon is often referred to as 'autogenous healing', and is referred to again in Chapter 4. A third is the slight expansion of concrete which occurs from what is sometimes called 'reverse drying shrinkage' or moisture movement of the concrete itself. A typical example is the walls and floor of a water retaining structure after it is filled and put into operation. The author's opinion is that it is generally advisable to seal all cracks in external members or members exposed to an aggressive environment, which are wider than 0·10 mm. If this work is properly carried out it should prevent the widening of the crack with the passage of time and thus avoid corrosion of the reinforcement and further deterioration.

Fine hair-line cracks in white concrete members, particularly when the surface is smooth, become a channel and collecting place for dirt, and after some years are very conspicuous, to the annoyance of the owner and embarrassment of the architect and contractor.

When advice is sought on cracks in reinforced concrete, the first step is to try and ascertain the cause of the cracking and to decide whether it is structural or non-structural. The request for an investigation may be made during or very soon after the construction of the building or it may be many years later. The sooner the investigation is started after the cracking is noticed the easier it will be to diagnose its cause; also the cost of remedial work will increase with time.

3.2.1 The Causes of Cracking

There are many causes of cracking in concrete structures, but for practical purposes these can be placed in three main categories:

1. Structural cracking. This would indicate that the stability and/or safety factor of the structure or part of the structure has been affected, but not necessarily to a dangerous level. Structural cracking can be caused by: (a) error in design; (b) the loading of structure in excess of the design load, due to change in use; (c) error or short-coming in the method of construction and/or the materials

used; (d) unforeseen accident, such as explosion, impact etc. (fire damage is placed in a separate category below).

2. Cracking due to fire. This may be partly structural and partly non-structural and is always accompanied by spalling and other damage.
3. Non-structural cracking. This is cracking resulting from causes other than those set out in 1 and 2 above. It can be placed in a number of main types: (a) plastic cracking, of which there are two kinds; (b) thermal contraction cracking in the early life of the concrete; (c) drying shrinkage cracking; (d) cracking caused by rusting of reinforcement.

A careful examination of the cracks will give valuable information and will generally indicate their cause. Such factors as their position, pattern, direction (vertical, horizontal or inclined), their direction in relation to the main reinforcement, *i.e.* whether parallel or transverse, are all important. Any deflexion of the member is also significant. The width of the cracks should be measured as accurately as possible, but this will vary appreciably as the crack winds its way around pieces of aggregate; a high degree of precision is neither necessary nor practical. The depth of cover to the reinforcement must also be measured.

If a reasonable assessment of the cause of the cracking cannot be made on the basis of the information obtained above, then the reinforcement drawings, calculations and the specification should be examined.

The author's experience is that the majority of cracking is non-structural and in existing buildings is caused by poor quality concrete and/or inadequate cover to the reinforcement. Plastic cracking appears very soon after the concrete has been cast. Thermal contraction cracking is usually present when the formwork is removed, or if there is no formwork, within the first 48 h of casting, although it is often not noticed for several days or even weeks. The application of a curing membrane helps to mask these fine cracks. The width of thermal contraction cracks is usually increased by normal drying shrinkage.

3.2.2 Non-Structural Cracking: Types, and Methods of Repair

Plastic Cracking

There are two categories of plastic cracking. The first, and most common, results from a too rapid evaporation of moisture from the surface of the concrete while the concrete is still plastic, and is usually referred to as plastic shrinkage cracking.

Investigations by various authorities have shown that the principal cause of plastic shrinkage cracking on horizontal surfaces is a rapid evaporation of moisture (drying out) from the surface of the concrete. When the rate of evaporation exceeds the rate at which water rises to the surface (known as 'bleeding'), plastic cracking is very likely to result.

The rate at which the water in the mixed concrete reaches the surface and the total quantity involved depends on many factors, all of which are not yet completely understood, but the following are known to play an important part in this phenomenon:

Grading, moisture content, absorption, and type of aggregate used
Total quantity of water in the mix
Cement content
Thickness of the concrete slab
Characteristics of any admixtures used
Degree of compaction obtained and therefore the density of the compacted concrete
Whether the formwork (or sub-base) on which the concrete was placed was dry or wet

The rate of evaporation from the surface will also depend on a number of factors which are much better understood, and these are:

Relative humidity
Temperature of the concrete
Temperature of the air
Wind velocity
Degree of exposure of the surface of the slab to the sun and wind

Plastic shrinkage cracking shows itself as fine cracks which are usually fairly straight and can vary in length from about 50–750 mm. They are often transverse in direction. In some cases several of the cracks are parallel to each other and the spacing can vary from about 50 mm–90 mm. The cracks are usually shallow and seldom penetrate below the top layer of reinforcement, although in severe cases they can extend to a greater depth and even right through the slab. They are generally numerous. Figure 3.1 shows typical plastic cracking in a floor slab.

It is quite common for this type of cracking to occur in hot sunny weather or on days when there is a strong drying wind, and can cause consternation for those who do not realise what it is. Unless it is very severe, when it may result in deep cracks and in a permanently weakened surface to the slab, it does no real harm. The cracks should be grouted in with a Portland cement grout well brushed in, and the treated surface should be covered with polythene sheets held down around the edges with planks and blocks for at least 48 h.

It has been found that the use of an air-entraining admixture to give a total air content of the concrete of $4\frac{1}{2} \pm 1\frac{1}{2}\%$ will usually help considerably in reducing the incidence of plastic cracking. Prevention is always better than

FIG. 3.1. Plastic shrinkage cracking in R.C. slab (courtesy: G. F. Blackledge).

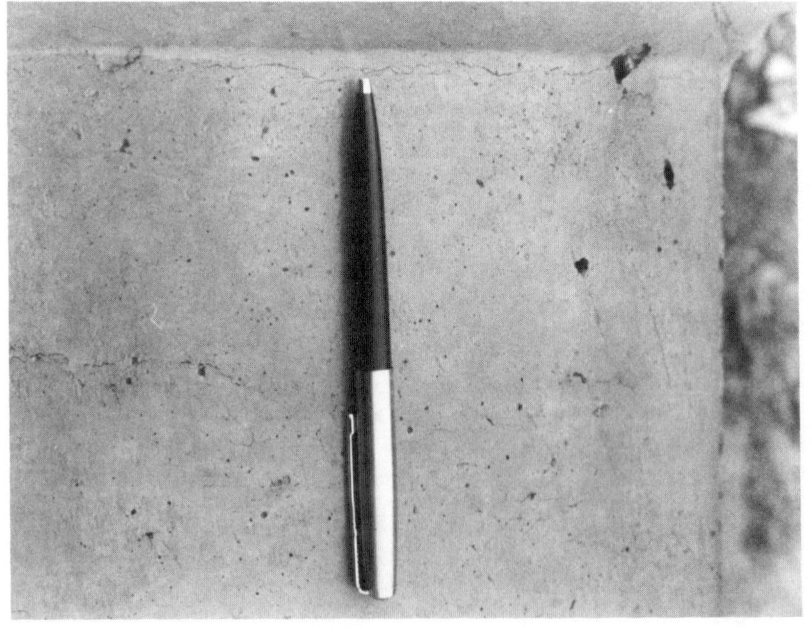

FIG. 3.2. Plastic settlement cracking at top of mushroom-headed column.

cure, and if the concrete is covered with plastic sheeting well lapped and held down around the perimeter, immediately finishing operations are complete, it is unlikely that plastic shrinkage cracking will occur.

The second form of plastic cracking is due to settlement and is quite different in its origin to plastic shrinkage cracking which has just been described. It can be caused in two basic ways. The first is by the resistance of

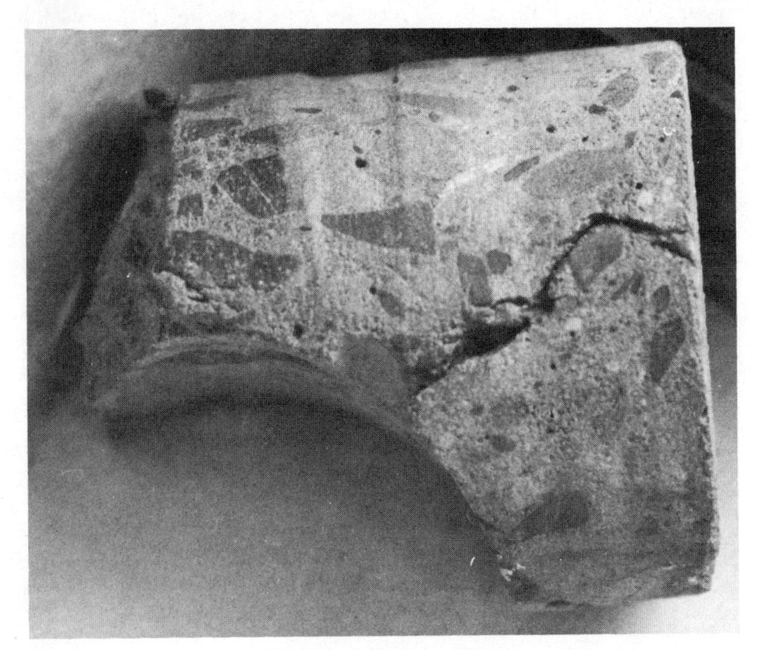

FIG. 3.3. Detail of plastic settlement crack.

the surface of the formwork to the downward settlement (compaction) of the plastic concrete under the influence of poker vibrators and the force of gravity. This resistance delays the movement, and when this does occur and the concrete has stiffened, a crack is very likely to form. This is invariably a surface defect and the cracks are wider at the surface and do not penetrate deeper than about 20–25 mm. The second is more serious, because the cracks penetrate at least to the reinforcement, they may be wider inside the concrete than on the surface, and may be associated with some degree of honeycombing. The cracks are caused by the concrete becoming 'hung up' on either the reinforcement or on spacers, and a crack forms as the concrete stiffens. Some changes in the mix proportions and greater care with compaction will generally solve the problem.

Figure 3.2 shows this type of cracking near the head of a mushroom-headed column. Figure 3.3 shows in detail how the crack tends to widen as it proceeds inwards. The author is indebted to H. Tomsett of the Cement and Concrete Association for these photographs.

It is advisable for cracks of this type to be repaired by crack injection as described in Part 2 of this chapter; simple surface treatment is unlikely to be adequate to ensure long term durability. Where this type of cracking occurs in deep beams and thick slabs, it is advisable to check whether there is any honeycombing in addition to the cracks, and remedial measures taken accordingly, as described in the section in this chapter on repairs to honeycombed concrete.

Thermal Contraction Cracking

During the setting and early hardening process of concrete considerable heat is evolved by the chemical reaction between the water and the cement which results in a rise in temperature of the concrete. The actual rise, the peak temperature, and the time taken to reach the peak and then to cool down, will depend on a large number of factors of which the following are the most important:

- The ambient air temperature
- The temperature of the concrete at the time of placing
- The type of formwork used (whether timber, plastic or steel), and the time the formwork is kept in position
- The ratio of the exposed surface area of the concrete, *i.e.* the area not protected by formwork, to the volume of concrete
- The thickness of the section cast
- The type of cement used and the cement content of the mix
- Whether any provision is made for the thermal insulation of the concrete after the formwork is removed
- The method of curing

The above are not placed in order of importance because this order depends on site conditions.

The detailed behaviour and characteristics of concrete as it hardens are not yet fully known and research on this is being carried out in the UK and other countries. As the temperature of the concrete rises the concrete expands and when it cools down it contracts. The coefficient of thermal expansion (and contraction), depends on a number of factors of which the principal are the type of aggregate and the mix proportions.

Unless the section (floor, wall or roof) is completely unrestrained (a state of affairs never met in practice), thermal stress will be developed as the concrete cools down and contracts. The greater the restraint the higher the thermal contraction stress. These stresses are generally tensile, but in certain parts of a structural member they may be compressive. These tensile stresses often exceed the strength of the concrete in tension and/or the bond strength between the concrete and the reinforcement, and then cracking will occur.

FIG. 3.4. Thermal contraction cracks in a balcony slab (courtesy: Professor B. P. Hughes, Birmingham University).

Thermal contraction cracks extend right through the member. Figure 3.4 shows typical cracks in a cantilever balcony slab. While these cracks are seldom significant structurally, they do form permanent planes of weakness through the member unless they are properly repaired. Normal drying shrinkage will tend to widen these cracks, which when first formed are usually very fine, often no wider than 0·05 mm. For this reason they are often not noticed for several weeks after casting. This type of cracking is often wrongly attributed to drying shrinkage. The latter takes place slowly under normal conditions in the UK and in the first 28 days after casting is unlikely to exceed about 25 % of the maximum which will eventually occur, over a long period of time.

The method of repair will depend on whether it is anticipated that movement across the crack will take place at a later date, that is, whether the crack will remain 'alive'. If no further movement of this kind is likely to occur, then the crack can be sealed with a rigid material, otherwise some degree of flexibility should be introduced into the repair. Exactly how this is done will depend on the environmental conditions and the type of finish to the member which is acceptable to the client. The choice lies between crack injection and surface treatment as described in Chapter 4 for water retaining structures, combined with a high quality flexible sealing and decorative coating. Crack injection is described in Part 2 of this chapter.

Drying Shrinkage Cracking

The author's experience is that drying shrinkage cracking is generally confined to (a) non-structural members which have either no reinforcement or only 'nominal' reinforcement for handling purposes; (b) thin toppings, screeds and rendering.

In most cases it is caused by a badly designed mix, the effects of which are aggravated by inadequate curing. The use of calcium chloride as an admixture, or the presence of chlorides in the aggregates, will increase drying shrinkage.

The inadequacies in the mix design may be just too much water in the mix, or the use of poorly graded fine aggregate which contains a high proportion of very fine material. The higher the percentage of fine material in a concrete or mortar, the higher will be the water demand to obtain a given workability.

All concrete and mortar shrinks on drying out, and as stated in the previous section, drying shrinkage will tend to widen cracks which have been caused by other factors such as thermal contraction.

It is important to remember that contraction due to drying shrinkage after 28 days is about 25% of that which will take place in 180 days.

The method of repair will depend on the circumstances of each case. With toppings, screeds and rendering, the drying shrinkage cracks may be accompanied by curling and de-bonding. In other cases, the extent of the cracking and the final surface finish will determine the remedial treatment.

Detailed recommendations for the repair of concrete floors are given later in this chapter.

3.2.3 The Repair of Non-Structural Cracks

The term 'non-structural cracking' is intended to cover cracks in concrete members where the accepted factor of safety is not endangered and where strengthening of the member by the addition of reinforcement and/or concrete is not required.

As previously stated, cracks as such are not necessarily detrimental to concrete. It is the reason for the formation of the cracks, their width and location and the condition of exposure of the member which are the deciding factors when considering whether the cracks should be repaired or left alone, and how the repairs should be carried out.

There is little basic difficulty in repairing non-structural cracks in building structures. However, a real problem can arise when it is desired to repair the cracks in such a way that the repair cannot be seen (or is unnoticeable) after completion. Unless the cracks are some distance from the nearest viewing point, it is virtually impossible to effect an invisible repair unless a decorative coating is applied over the whole of the member on completion of the work.

It has been noted earlier in this chapter that cracks in the exposed faces of external members are likely to become gradually wider and more conspicuous with the passage of time. This applies particularly under conditions of extreme exposure, and with light coloured concrete in a town atmosphere.

The cracks being considered here do not include 'crazing' and 'map cracking'. Crazing is defined in British Standard, BS 2787: Glossary of Terms for Concrete and Reinforced Concrete, as 'the cracking of a surface into small irregularly shaped contiguous areas'.

As previously stated cracks in external members and membranes in an aggressive environment which are wider than 0·10 mm should be sealed. If there are no rust stains and no spalling of the concrete, and tapping with a hammer along the line of the crack does not reveal any hollowness, it is reasonable to assume that no significant corrosion of the reinforcement has occurred. In such a case, cutting out of the crack in order to repair it is not recommended. It may be decided to cut out the concrete in a few places to check the condition of the reinforcement, but that is all. Also, it may be decided to take some samples of the concrete to check the quality etc. and determine the chloride concentration. The recommendations given below can be used to repair shallow cracks due to 'miscellaneous non-structural causes'.

When appearance is not important, careful tapping along the line of the crack with a cold chisel is recommended. This will reveal any slight honeycombing. It is emphasised that this is not a cutting out of the crack. All grit, dust and dirt should then be brushed out, and the surface of the concrete on each side of the crack should be brushed with a wire brush. This preparation should be followed by the brushing in of a latex grout made with two parts Portland cement and one part of styrene butadiene latex emulsion by weight. The grout can with advantage be brushed into the surface each side of the crack for a distance of about 75 mm, *i.e.* the area which has been wire brushed. If it is practical, a further brush coat of grout can be applied one to two weeks later.

For concrete members which are exposed to view, the following method of repair is suggested. It is emphasised that this applies to fine cracks where there is no corrosion of the reinforcement and no spalling of the concrete.

The first step is to clean with cold water the concrete on both sides of the cracks for a distance of about 75 mm and then insert by means of a squeegee, a thin mortar or grout made with either white cement or a mixture of white and grey cement so as to obtain the best possible match with the rest of the concrete. The addition of a white artificial rubber latex will give beneficial results by reducing permeability and shrinkage. Not less than two weeks after the completion of the repair, the whole of the concrete members should be washed down with water. The whole facade should then weather fairly evenly.

The author is generally in favour of suitable decorative waterproof coating applied as a finish to major repair work, but these coatings do introduce a maintenance commitment which some architects and building owners are reluctant to accept. It should be remembered when considering this problem that concrete which has been repaired is not really as good nor as durable as similar quality concrete which has never needed repair. The complete sealing of the whole area as a final finish to the repair work is in most cases desirable and worthwhile.

3.3 GENERAL REPAIRS

The use of poor quality concrete, even if the cover to the steel complies with what may be termed 'standard requirements', is likely to result in corrosion of the reinforcement on the external face(s) of the member.

British Standard Code of Practice, CP 110: The Structural Use of Concrete, relates cover to reinforcement to degree of exposure and quality of the concrete. The details are given in Clause 3.11.2 and Table 19 of the Code. Prior to the issue of CP 110, the relevant Codes were CP 114: The Structural Use of Reinforced Concrete in Buildings, and CP 116: The Structural Use of Precast Concrete. These Codes are still in force for those who wish to use them. It is perhaps relevant to remark that many engineers would stop using CP 114, were it not for what they consider are the unnecessarily complicated design requirements in CP 110.

Under CP 114, the recommended cover to reinforcement was not related to the degree of exposure nor to the quality of the concrete except where the concrete was in direct contact with the ground or was exposed to particularly corrosive conditions (Clause 307 of the Code).

There can be no disagreement with the basic premise in Clause 3.11.2 of CP 110 that the depth of cover should be related to the quality of the concrete. The author has seen examples of reinforcement protected with a depth of cover 20–25 mm which have been exposed to marine conditions for over 30 years without any attack developing on the steel. This is referred to in more detail in Chapter 5.

Apart from the quality of the concrete, the important factor in concrete members exposed to conditions of severe exposure is to ensure, as far as this is practical, that physical damage does not occur and that cracks do not develop. Where reinforcement is used to control cracking, the closer this is to the surface, the better the crack control; this is the principle used in ferrocement.

What is needed is an impermeable concrete of adequate strength which is free of cracks which exceed an 'acceptable' width. Detailed consideration has been given to cracking and crack widths in the previous section.

The general procedure for carrying out repairs to spalled, cracked and generally deteriorated concrete members, excluding special crack treatment, is as follows. All defective concrete is removed by hand or pneumatic tools, or by high velocity water jets. The use of high velocity water for cutting and cleaning concrete has only recently been introduced in this country. It is unfortunate that the construction industry is so slow and reluctant to adopt new ideas. This technique has a great deal to recommend it particularly as it can be used for cleaning the reinforcement as well. The pressure at the nozzle is between 200 and 650 atm (3000 to 10 000 lb/in^2). Comparatively little water is used per jet, about 50 litres/min; of this quantity, about one-third is dissipated as mist and spray.

The jets leave the concrete and the steel clean and damp, and in this way corrosion products on the steel are removed which does not occur with other methods of steel cleaning. A very thin film of iron oxide will form quickly on the cleaned steel, but experience shows that this film tends to protect the steel and it does no harm.

Other advantages of using high pressure water is that there is no vibration and practically no noise except that from the motor and pump, and the speed of working is considerably higher than when using percussion tools. Further information on this technique is given in Chapter 5.

The spalled and defective concrete should be removed back to sound concrete, and to the extent required to remove rust from the reinforcement. There is no need to go further than this. Figure 3.5 shows spalled concrete being removed prior to cleaning of reinforcement, which was followed by application of gunite.

After the completion of the above work and immediately before the application of the new mortar or concrete, the reinforcement and the surrounding concrete should be given a coat of Portland cement grout. The grout should be made of two parts of OPC to one part of a styrene butadiene or acrylic latex, by weight. This should be well brushed into the concrete and applied evenly to the reinforcement. The object of the grout is to create an intense alkaline environment around the steel and to improve bond between the old concrete and the new mortar or new concrete. The grout should not be allowed to set before the mortar or concrete is placed in position; a maximum period of 20 min should be allowed between the two operations. The author is not in favour of the use of acid rust inhibitors; these are usually based on phosphoric acid. The use of acids in contact with

FIG. 3.5. Spalled concrete being removed prior to cleaning reinforcement (courtesy: Cement Gun Co. Ltd).

concrete should be avoided and in any case the real passivating value of these compounds is open to question. Contractors who use them claim that the acid is neutralised by the addition of a subsequent coat of material with a high pH, but the author feels it is better not to use the acid at all. The material used for the repair has been described as 'mortar or concrete'. The reason for this is that both materials can be used successfully, but it is only practical to use concrete in positions where it can be properly compacted. This is one of the reasons why most repairs are carried out with a mortar.

Except for small repairs to chipped or broken arrisses the author favours the use of a Portland cement mortar, rather than a resin mortar. In this context, resin mortar means a mixture of an organic polymer and fine aggregate, in other words the cement is replaced by the polymer. These resin mortars have their uses for special purposes, but are very expensive compared with the same volume of cement/sand mortar, and for general repairs to building structures the extra cost is not justified. When such an expensive material is being used there is a tendency for the contractor to reduce the quantity required to a minimum and this can lead to inadequate removal of the old defective concrete. Another factor is that resin mortars are generally very much stronger than normal concrete and this extra strength is not required, and is in fact undesirable. The strength of the new mortar should be compatible with the strength of the base concrete to which it is bonded.

This question of compatibility of strength between the old and new work assumes considerable importance when the concrete to be repaired is itself of rather poor quality. If the new mortar or concrete is very much stronger than existing concrete it will tend to pull itself away from the background. In some cases it may be necessary to use light galvanised mesh pinned into the base concrete with stainless steel pins or wired to stainless steel bolts which have been securely fixed into the concrete.

Two of the most important factors in a satisfactory repair are: (a) to ensure the best possible bond between the old and new work, and (b) to ensure that the new concrete or mortar is as dense and as impermeable as possible, compatible with its mix proportions.

All architects, engineers and contractors know that concrete must be properly compacted, but how many realise that the same applies to cement/sand mortar used for rendering and floor screeds, and for repairs? Other things being equal, the quality of the final product will be directly related to the amount of compactive effort put into the application of the mortar repairs. A properly graded clean sand is essential. The particle size, *i.e.* the percentages retained on the standard sieves, will be governed by the method of application of the mortar. For guniting (pneumatically applied mortar at relatively high velocity), a carefully graded concreting sand to BS 882 will give the best results, while for hand applied mortar, the coarser gradings of the building sands to BS 1198–1200 should be used.

With hand application, unless considerable pressure is exerted with the trowel, hollow spots may develop as the mortar matures and dries out. No hollow areas at all should be accepted in this type of repair work.

Figures 3.6, 3.7 and 3.8 show 'before', 'during' and 'after' repairs to a high rise building.

About one month after the completion of the repairs it is usually noticed that fine hair cracks have formed around the perimeter of the new areas of concrete or mortar. These are shrinkage cracks and unless steps are taken to seal them they are likely to form the nucleus of fresh deterioration. The cracks will tend to widen in the course of time, perhaps several years, and then moisture will gain access to the reinforcement and rusting and spalling will start again. For this reason the author considers it prudent to seal the perimeter of the repaired areas with two brush coats of grout consisting of two parts OPC and one part styrene butadiene or acrylic latex. Some specialist contractors offer this final sealing combined with a durable decorative treatment. It is obvious that no matter how much care has been taken in matching the new to the old work, the surface of the repaired

FIG. 3.6. Deteriorated reinforced concrete columns in multi-storey block, prior to repair (courtesy Gunac Ltd).

FIG. 3.7. Deteriorated columns in Fig. 3.6 prepared for application of new protective pneumatically applied mortar (courtesy: Gunac Ltd).

FIG. 3.8. Concrete columns in Figs. 3.6 and 3.7 after repair but prior to final treatment (courtesy: Gunac Ltd).

members will have a very patchy appearance. Some form of final decoration is therefore desirable, although many clients object in principle to this on the grounds that this will introduce a recurring maintenance expense. This is no doubt correct, but the author cannot see any technically satisfactory alternative.

Figure 3.9 shows a high rise building after final decorative treatment following extensive repair work.

The information and recommendations given may suggest that the work is straightforward and relatively simple. However, in practice, many problems arise which must be correctly solved if a satisfactory job is to result. These problems include the following:

The selection of the materials, *i.e.* resin mortar or cement-based mortar
How much concrete should be removed and the best method for removing it, and cleaning the reinforcement
The mix proportions, and if cement mortar or concrete is used, the grading of the aggregates and the w/c ratio; the use of admixtures such as latex emulsions
How should the mortar be applied, *i.e.* by hand or compressed air
The selection of the materials for final sealing and decoration

Repair work is expensive, but if properly executed should give a maintenance-free life of 10–15 years, unless there is an undesirable concentration of chlorides in the base concrete. This latter condition is discussed later in this chapter.

3.4 REPAIR OF HONEYCOMBED AND UNDER-COMPACTED CONCRETE

The previous sections have dealt with the principles of general repairs to concrete members where the defects are usually only too obvious. However, cases occur where the defects are invisible, such as inadequate compaction and honeycombing of the concrete within a member, such as a beam, column or wall. There may be some signs on the surface which to an experienced eye suggest that all may not be well in the interior of the member. Factors which are likely to cause this type of trouble include congested reinforcement, low slump concrete, breakdown in poker vibrators, premature stiffening of the concrete as it is being placed, movement of the formwork.

The suspected concrete should be checked by an ultrasonic pulse velocity survey (upv), gamma radiography, or by taking cores. If this investigation demonstrates that the concrete is defective then it either has to be repaired or removed; this latter course may involve demolition of the member. Demolition is difficult and expensive and can cause serious delay in the

Fig. 3.9. Concrete columns in Fig. 3.8 after final sealing/decorative treatment (courtesy: Gunac Ltd).

completion of the structure, and in fact is seldom justified. The removal of the honeycombed or under-compacted concrete and its replacement by new concrete is usually a satisfactory solution. The major difficulty in such work is to ensure that the new concrete is itself fully compacted. The author has found that the procedure described below, when carefully carried out, will give satisfactory results. This refers to the repair of a load bearing wall or column where the defects are near the base, but the principles can be adapted for other members.

The concrete must be cut away to expose the under-compacted and honeycombed area. It may be prudent to provide temporary support to the member while the repair is in progress.

When the extent of the honeycombing has been determined by such work, proceed to remove all sub-standard concrete. The cutting away can be carried out by percussion tools, thermic lance, or high velocity water jets, whichever is more appropriate. Percussion tools produce noise and vibration. Thermic lances may damage the reinforcement, but do not produce noise and vibration. Information on the use of high velocity water jets is given in Chapter 5. Whichever method is used, sufficient concrete must be cut away to provide adequate space for placing and compacting the new concrete. Figure 4.6 (p. 137) shows this type of work for the wall of a water-retaining structure.

Even under the most favourable conditions, compaction of the concrete is likely to be difficult, and unless all precautions and great care are taken, it may well be impossible. The essentials are a cement-rich mix, well graded aggregates, low w/c ratio and sufficient workability for proper compaction. This will certainly require the use of a plasticiser. Consideration should be given to the use of one of the new superplasticising admixtures. Some brief information on these new materials is given in Chapter 1.

It is not suggested that the above will necessarily result in the repaired member having the same strength as the original was intended to have. However, if allowance is made for redistribution of stress between the old and new concrete and the reinforcement, and provided the work has been properly carried out, the small reduction in strength should be insignificant.

For honeycombing in non-loadbearing members, pressure grouting with a cement grout, executed by a specialist firm, should be adequate. Even so, because of the cost of the equipment needed for this work, for small jobs, cutting out and repair may be more economic. Pressure grouting is referred to in more detail in Chapter 4.

3.5 THE REPAIR OF BINS AND HOPPERS HOLDING COARSE-GRAINED MATERIALS

Two materials in general use in industry and which are stored in open bins and hoppers, are limestone and coke. Both of these can cause severe abrasion to the inside of the structure but limestone has no adverse effect on the concrete. On the other hand, coke contains chemical compounds and when it is slaked with water the resulting effluent may be very aggressive to Portland cement, due principally to the presence of mineral acids and sulphates in solution.

The deterioration of concrete in these structures usually follows a particular pattern. Inside there is abrasion, particularly on the surface of the sides and bottom of the hopper, and in the case of coke holding bins, there is likely to be chemical attack as well. With older structures, rusting of reinforcement, cracking and spalling of the concrete may occur both on the inside and externally. This is largely due to rather poor quality concrete, and inadequate cover to the reinforcement. To this is sometimes added structural distress resulting from underestimation in the original design of the actual stresses developed during filling and emptying.

The logical, and best material for the repair of these structures is high quality reinforced gunite. High quality gunite is dense and impermeable and very resistant to abrasion. These are the qualities in mortar and concrete which form the first line of defence against chemical attack.

The use of sulphate-resisting Portland cement is generally effective in preventing deterioration of the concrete by sulphates in solution (except ammonium sulphate), up to a concentration of about 5000 ppm. Above this figure, and where acids are present, consideration should be given to the use of high alumina cement. One of the basic requirements in the satisfactory performance of HAC concrete and mortar is a low initial w/c ratio. This is met by high quality gunite which has a w/c of about 0·35. Further information on HAC is given in Chapter 1. Abrasion resistance of cementitious materials is directly related to compressive strength. The 28 days equivalent cube strength of good quality gunite should be in the range of 55–65 N/mm^2.

In a number of cases it has been found that the slope provided for the hopper bottom is insufficient and this causes trouble with emptying the bin. To solve this problem it may be necessary to increase the slope to something over 50°, and to do this in concrete requires a top shutter. Placing and compacting concrete on such a slope with a top shutter is very difficult and the result can be disappointing to say the least. If concrete is used it can be finished about 100 mm below the required finished level and then a layer of reinforced gunite 100 mm thick can be applied. An alternative is to construct a 'false' bottom in heavily reinforced gunite. This method was adopted for limestone hoppers in a cement factory.

As with all work involving bonded toppings, careful preparation of the base concrete is essential. This involves the removal of all dirty and deteriorated concrete and the cutting out of all major cracks. Further details of this type of work are given in Part 2 of this chapter.

3.6 THE WEATHERING OF CONCRETE BUILDING STRUCTURES

The effect of weathering on a concrete structure depends on the type and quality of the concrete and the degree of exposure. In the vast majority of cases, repairs to the structure are not required as a result of weathering alone. In fact in country districts weathering can result in a marked improvement in the appearance of the concrete without causing any deterioration at all.

Weathering may be defined as changes in the characteristics of the material, and particularly in its surface appearance, brought about by the weather. The author would not include the carbonation of the surface layers of the concrete in the general definition of weathering. Some basic information on carbonation has been given in Chapter 2.

Apart from carbonation which is caused by the presence of carbon dioxide in the air, the causes of weathering are numerous and include the following:

(a) Dirt in the atmosphere deposited on the surface of the concrete by wind, rain and fog
(b) Deposits washed onto the surface from adjoining surfaces
(c) Chemical attack from aggressive compounds in the atmosphere
(d) Fungal and algal growths on the surface
(e) The general effect of wind, rain, and atmospheric temperature changes

The deposits formed under (a), (b) and (d) may in the long term provide a protective coating against attack from (c).

Concrete surfaces, particularly plain ones, are very vulnerable to weather staining. This is due to a number of factors, including the fact that the surface of concrete is absorbent and that it is usually cast in comparatively large light coloured areas. Brick and stone are also absorbent (more so than concrete), but the units are very much smaller and each small area is separated from adjoining ones by joints made of different material and having a different texture and colour tone.

The inherent problems associated with the weather staining of plain concrete, is the main reason for the move away in recent years, from this type of surface, to that of exposed aggregate and moulded finishes. Figure 3.10 shows weather staining of concrete on a new office building in London. Weather staining in itself is not detrimental to concrete and it is only when there is chemical attack from atmospheric pollution that damage occurs. As stated above, a thick layer of grime is likely to help prevent or reduce attack. However, in a chemically aggressive environment, as exists in some industrial towns, when the building is cleaned, the newly exposed surface of the concrete or cast stone will be exposed to intensified attack.

Smoke pollution has been substantially reduced during the past 15 years in many of the large cities in the UK. In London this reduction is particularly noticeable and it is therefore all the more unfortunate that the level of sulphur dioxide (SO_2) has not decreased and is now probably higher than in any other town in the country. When sulphur dioxide is dissolved in rain water or fog, sulphurous and sulphuric acids are formed which are very aggressive to Portland cement concrete, and calcareous materials such as limestone.

Fig. 3.10. Weather staining on precast concrete units in office block in London (courtesy: Cement and Concrete Association).

It is therefore advisable to give these facts careful consideration before embarking on a scheme of external cleaning. The cleaning will certainly improve the appearance of the concrete but may result in accelerated deterioration, although this is likely to apply more to certain types of stone masonry than Portland cement concrete. Where it is feared that deterioration may occur, it is difficult to put forward recommendations which will not alter to some extent, the light reflecting properties of the surface. Also, all applied coatings undergo some degree of colour change. Ideally, this type of coating should be colourless, of low viscosity, very durable, bond well to the substrate, and should not lose its transparency. The author knows of no material which fulfils all these conditions entirely satisfactorily, and so some compromise is required.

Silicones probably give the best results from the point of view of appearance, but are short-lived. Orthosilicates have an appreciably longer life, but are rather more visible than silicones. Low viscosity epoxide resins are very durable but are likely to change the appearance of the concrete or cast stone.

However, the application of waterproofing liquids should be carried out with care, particularly if it is not intended to treat complete wall areas. This is referred to again in more detail later in this chapter under the waterproofing of walls.

There is another type of weathering which can cause deterioration of concrete and cast stone, and this is frost action. This is unlikely to occur in the UK if good quality concrete and cast stone are used, but poor quality materials can be damaged in this way on exposed sites. With porous material, such as poorly compacted concrete, and concrete and cast stone made with a high w/c ratio, moisture penetrates the surface layers and when it freezes it expands and the expansion causes spalling. It should be particularly noted that high quality concrete is much more impermeable, and because of its higher strength is better able to resist the stresses caused by the expansion of absorbed moisture when this freezes. On very exposed sites, it is therefore advisable to use air-entrained concrete. This type of concrete is now recommended for all external paving as it will successfully resist the disruptive effects of de-icing salts. Air-entrained concrete and mortar should not be confused with aerated concrete and mortar which are quite different materials. Brief information on air-entraining agents has been given in Chapter 1.

The air-entraining resin used must be carefully dispensed into the mixing water so as to ensure uniform distribution throughout the batch of concrete or mortar. The dosage must be such that the total air content of the mixed material is in the range of $4\frac{1}{2}\% \pm 1\frac{1}{2}\%$. The entrainment of this amount of air has the effect of reducing the compressive strength of the concrete by about 10%, but at the same time it acts as a workability aid and thus the w/c ratio can be reduced, which in turn will increase the strength of the concrete. In repair work, impermeability and durability is usually of greater importance than strength and so the small net reduction in strength is not important.

3.7 WATERPROOFING CONCRETE BLOCK AND PRECAST PANEL WALLS

It is standard practice in the UK for the external walls of buildings to be constructed as cavity walls, the cavity being about 50 mm wide, and the inner and outer leaves being brick or blockwork. With brick and block cavity walls, there are standard details for ensuring watertightness at door and window openings. In addition, it is essential for the wall ties to be kept clean, *i.e.* free from mortar droppings. The principal cause of moisture penetration through these walls is failure to observe these basic requirements (water seals at openings and clean wall ties).

Due to the 'energy crisis' and the need to conserve fuel, there has been considerable pressure to improve the thermal insulation of existing buildings, particularly the walls. One solution, the proposal to fill the 50 mm wide cavity with material of high insulating value, sparked off a considerable technical argument. Some authorities claimed that the filling of the cavity would give rise to serious problems of moisture penetration, while others felt these fears were exaggerated. At first, cavity fill required individual relaxation under the Building Regulations, but now under a 'type relaxation' the work can go ahead when it is covered by an Agrément certificate.

For cavity fill to be successful it is important that the external leaf of the wall does not permit ingress of rain under prevailing weather conditions. It is recommended that owner/occupiers should seek advice from such an organisation as the Building Research Establishment before embarking on the filling of the cavity in an external wall. The materials of which the outer leaf is built and the degree of exposure are vitally important matters; if these factors are wrongly interpreted, damp walls are likely to result.

With the advent of industrialised building in the 1960s, precast concrete panels came into use. In many cases these panels are of sandwich construction. The outer skin is normal dense concrete, while the inner skin is often lightweight concrete for thermal insulating purposes. The narrow gap between the two leaves is about 20 mm wide and may be filled with expanded polystyrene or similar material.

With precast concrete panels it is important that the vertical and horizontal joints between the panels should be correctly detailed and sealed; door and window openings should also be sealed to prevent ingress of moisture, even though the cavity is only 20 mm wide and is filled with an insulating material. In some proprietary systems, the cavity face of the external leaf is sealed with a sheet membrane. This will provide a substantial measure of resistance to the penetration of moisture from the outside. However, it should not be completely relied upon to the exclusion or neglect of standard details for sealing around door and window openings. The standard 'tray' used in normal cavity construction has proved its value over a long period (more than 50 years), but proprietary systems used by the manufacturers of precast panels have had to be modified several times during the past 15 years. Even now, many problems of water penetration into high and medium rise buildings are constantly coming to light. The joints between the panels may be sealed with an insitu flexible material or with a loose 'tongue'. Unfortunately, some of these materials have not proved particularly durable. Probably the most durable material at present on the market is preformed neoprene. The use of a preformed material for sealing the joint requires close tolerances on the width of the joints and smooth and accurately formed sides to the joints. These cellular gaskets are compressed and inserted into the joint with a primer or adhesive.

Problems arise if cracks, which are usually quite fine, develop in the block, bricks, or panel walls, particularly if this is accompanied by areas of porosity. The author is not in favour of trying to waterproof limited areas of the external surface of a wall. The reason is that flow of rain water down the waterproofed surface will be appreciably greater than from the untreated areas. This can result in dampness appearing inside the building at a lower level because the wall below the treated area receives much more water than it did before. Concrete and clay bricks are to a greater or lesser degree absorbent, which means that when rain is driven against a wall some of it is absorbed and the remainder runs off mostly in a downward direction.

Various proprietary preparations based on silicones are on the market and are used for waterproofing walls. These are fairly short-lived materials and form a very thin water-repellent skin on the surface of the wall. They are effective, subject to what has been said above about the risks of partial treatment, provided the wall surface, *e.g.* rendering, does not develop further cracks after treatment.

For the long term effective solution of water penetration of walls, the best solution is generally careful initial preparation (filling cracks, removal and replacement of unbonded rendering etc.) followed by the application of a specially formulated waterproof decorative coating.

3.8 REPAIRS TO RECONSTRUCTED AND CAST STONE

The repair of reconstructed or cast stone, which is in fact a type of precast concrete, presents certain special problems. Cast stone products are covered by BS 1217: 1975: Cast Stone. BS 2847: Glossary of Terms for Stone used in Building, defines cast and reconstructed stone as 'A building material manufactured from cement and natural aggregates for use in a manner similar to and for the same purpose as natural building stone'. The definition of cast stone in BS 1217 is similar, but not identical to the above.

This material is used as a finish in its own right and therefore the 'special problems' arise from the general characteristics of cast stone. British Standard BS 1217: 1975 lays down certain requirements, *e.g.* 30 mm minimum cover to untreated mild or high tensile steel reinforcement, while non-ferrous stainless steel and galvanised steel must have a minimum cover of 10 mm. The minimum compressive strength of the cast stone is 25 N/mm^2. BS 1217 includes provision for an Initial Surface Absorption Test (Clause 11). This test (ISAT) is described in BS 1881: Part 5 and illustrated in Figs. 7 and 8 of that Standard. These changes will all help considerably to improve the durability of the product.

In spite of the tightening up of the requirements in the revised Standard, the author considers it is advisable for specifiers of new cast stone to insist on all reinforcement being hot-dip galvanised or given other equally effective protection.

Cast stone is often used for sills, copings, window surrounds and decorative panels. These are all non-structural elements, but are usually provided with 'nominal' reinforcement for handling purposes. All these elements are of small section, especially sills and window surrounds, and

FIG. 3.11. View of deteriorated cast stone sills.

therefore the cover to the reinforcement is reduced to the minimum and often below this. The reduced cover, combined with what unfortunately in the past has frequently been a material which is little better than a porous mortar, leads inevitably to the corrosion of the reinforcement. The corrosion causes cracking and spalling of the units. In an aggressive industrial atmosphere, there may also be some chemical attack on the cement matrix. In some cases the units have been found to be carbonated to a depth of 25 mm in a period of 10 years. Further information on carbonation is given in Chapter 2. Figure 3.11 shows deteriorated cast stone window sills.

To obtain satisfactory results when carrying out repairs, it is advisable to obtain the same type of aggregate as was used in the original units. Even if this can be done, complete harmonisation of colour and texture cannot be achieved, mainly because of the effect of weathering on the original units. A number of trial mixes and patching with these in selected areas is essential. The mix used should not only be compatible in colour and texture but also in strength. If a strong cement–sand mortar is used to repair a fairly weak and porous substrate the new work is likely to pull away from the old. This means that the new material should not be very different to the old which has proved to be unsatisfactory. The solution then lies in providing a complete waterproof decorative treatment to the whole of the cast stone units. A less expensive alternative is to apply two coats of a colourless waterproofing liquid of low viscosity to the whole units after repair. This will not mask the patching.

It can be seen that repairs to cast stone can be very difficult and in some cases it may be prudent, and more satisfactory and less costly in the long term to renew the units completely.

3.9 DETERIORATION OF CONCRETE DUE TO CALCIUM CHLORIDE IN THE MIX

Some information on the effect of using calcium chloride as an admixture in concrete, and the effect of concentrations of chlorides in mixing water and aggregates has been given in Chapter 2.

In reinforced concrete structures, the main trouble arising from the use of calcium chloride in the concrete is due to the fact that it corrodes the steel reinforcement and this in turn causes cracking and spalling of the concrete leading to increased corrosion.

Once the presence of chlorides (usually calcium chloride) has been established in concrete, the question naturally asked is whether any steps can be taken to neutralise its effects. At the present time, from a practical point of view, there is very little that can be done to ensure a satisfactory repair.

The maximum amount of anhydrous calcium chloride permitted in reinforced concrete is 1·5% by weight of cement. Even so, a large number of very strict conditions have to be complied with. These relate to the quality of the concrete and the depth of cover to the reinforcement. In cases where there has been corrosion of the reinforcement associated with the presence of calcium chloride, it is found that either these conditions have not been complied with or the dosage of calcium chloride was higher than 1·5% by weight of cement. However, when the concrete is poor quality or the cover to the steel is inadequate, concentrations appreciably lower than 1·5% can cause corrosion of the reinforcement.

When samples of concrete are analysed it is usual to find considerable variations in the chloride concentration, some samples having little or only a trace, while others have a high percentage. It has been claimed that chloride ions migrate from regions of high concentration to regions of low. There are no doubt special cases where migration does occur, such as along cracks and possibly through the more permeable parts of the concrete members. However, the author has found no clear information, based on research or controlled site trials, that is at all convincing, that migration on any significant scale does in fact occur through good quality dense concrete.

The actual execution of the repairs will be similar to that described in previous sections. Special care should be taken to reintroduce an intense alkaline environment around the steel reinforcement as this will have been broken down by the chloride ions in the concrete. The exposed and cleaned reinforcement can be coated with a cement grout using a styrene butadiene or acrylic latex emulsion as the gauging liquid. It is also recommended that a similar latex emulsion should be used in the mortar and concrete used for the repairs. It is important to exclude all moisture from the concrete, as far as this is practicable, by the application of a final sealing and decorative coating.

It must be stated however, that even when all this work has been carefully carried out, deterioration is liable to start again after some years. The unfortunate owner may thus be faced with a continuous programme of maintenance.

3.10 THE REPAIR OF CONCRETE ROOFS

Some of the special problems which arise when defects appear in concrete roofs of water-retaining structures are discussed in Chapter 4, and many of the recommendations given there are also applicable to the roofs of buildings.

One major difference however lies in the fact that most building structures, except multistorey car parks, are provided with thermal insulation and a vapour barrier, and this can complicate the locating and repair of the leaks.

Thermal insulation can be placed in three positions:

(a) Below the structural slab
(b) Directly on the structural slab and below the waterproof membrane
(c) On the structural slab, but above the waterproof membrane

It is the author's opinion that whenever possible, thermal insulation should be of a type which is not adversely affected by penetration of moisture, or which is impermeable to moisture. In the case of position (b), if the thermal insulation becomes saturated, it may be virtually impossible to dry it out. Work on this problem has been carried out by the Building Research Station at Garston, and at least one publication on the subject has been issued. In spite of this serious objection to the location of insulation, it is the position generally adopted in the UK.

Location (c) is known as the 'upside down' roof, and is popular on the Continent; the insulation must be impermeable to moisture. The disadvantage is that any defect in the membrane involves removing part of the insulation in order to effect a repair.

The membrane material used for roofs can be divided into two main categories: (1) sheet material; (2) insitu coatings and toppings. The sheet materials are built-up roofing felts, synthetic rubbers, and PVC. The insitu materials include mastic asphalt, and a wide range of coatings based on tar, bitumen and organic polymers. When mastic asphalt is used, an isolating membrane is laid on the concrete slab so that the asphalt does not bond to the concrete.

However, it is normal practice to bond the other insitu materials to the concrete deck as these are very thin compared with mastic asphalt. The minimum thickness recommended in CP 144: Part 2: Roof Coverings–Mastic Asphalt, is 20 mm for flat roofs and 25 mm for areas subjected to foot traffic. Bituminous compounds and organic polymers including a sprinkled wearing surface (when required), are seldom more than 2 mm thick.

Sheet materials can be fully bonded, partially bonded, or unbonded. The use of unbonded sheeting appears to have originated on the Continent where PVC sheeting is principally used.

From the above it can be seen that flat roof construction is rather complicated, and in fact there is one further complication namely the need for a vapour barrier between the concrete deck and any thermal insulation which may be adversely affected by moisture.

When considering leakage through a roof slab all the above must be taken into account which means that the repairs are likely to be difficult and expensive. A thorough knowledge of building construction, assisted by the architect's drawings is needed when investigating roof defects. Dampness in an external wall near the soffit of the roof deck may be due to a defect in the roof covering or the incorrect detailing of parapet walls. Useful information on this can be found in the Architect's Journal Information Library–Everyday Details.

Flat concrete roofs in the UK are subject to a wide temperature range, which depends to some extent on the colour of the surface, but can amount to 55 °C (from −5 °C to 50 °C). This range (for the structural slab) will be substantially reduced when the thermal insulation is placed on top of the slab. When the membrane is located above the insulation it will be exposed to this temperature range, but if fixed below, it will be protected not only from the extremes of temperature but also from ultraviolet radiation.

Roof slabs of multistorey car parks are not usually provided with thermal insulation, and in many cases the deck forms a parking area. Some designers omit a waterproof membrane and rely on the concrete slab for watertightness. In many such cases the result is unfortunate as water penetrates through fine hair cracks and small areas of slightly porous concrete. The water when it drips from the soffit of the slab is highly alkaline and damages the paintwork of cars parked below. There is seldom difficulty in locating the defects, but problems arise with regard to repairs.

The basic factor which must be taken into account in carrying out repairs is to allow for the thermal movement in the slab arising from the temperature range mentioned above. The author's experience is that it is useless to attempt repair with rigid materials. If the deck is used for parking, then this must also be allowed for in selecting the method and materials to be used for the repair. Unless it is possible to provide a complete membrane, adequately protected against damage, long term success may not be possible.

A full chapter could be devoted to the various alternative methods and techniques which could be used, and even then it would not cover all possible cases. As stated many times in this book, careful preparation of the base concrete, including removal of all dust and grit from cracks which have been cut out and from the surface of scabbled concrete is essential. The materials used for repair must be flexible and must bond well to damp concrete.

Figure 3.12 shows fully bonded PIB sheeting being laid on a concrete

FIG. 3.12. View of fully bonded PIB sheeting being laid on concrete roof (courtesy: Plysolene Ltd).

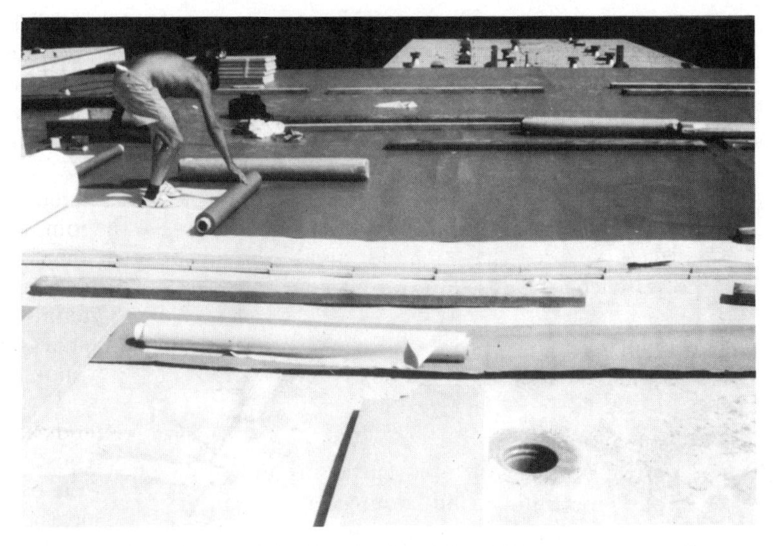

FIG. 3.13. View of unbonded PVC sheeting being laid on concrete roof (courtesy: Alkor Plastics (UK) Ltd).

roof. Figure 3.13 shows unbonded PVC sheeting being laid on a concrete roof.

3.11 THE REPAIR OF CONCRETE FLOORS

The repairs described in this section are intended to apply to industrial concrete floor slabs which have a concrete or cement-based wearing surface. Information and recommendations on the protection of concrete against chemical attack are given in Chapter 6.

There are in reality two basic categories of industrial floors, namely, lightly loaded and heavily loaded. While the principles of repairs are similar for both categories, the materials and techniques used may be different.

3.11.1 PATCHING AND GENERAL REPAIRS

Patching of numerous small areas may provide a perfectly satisfactory answer to a deteriorated concrete floor slab. In other cases a complete new topping is required, and this type of repair is dealt with under bonded and unbonded toppings. This patching and repair work should generally be carried out on the following lines:

(a) All weak and defective concrete should be cut away and all grit and dust removed. The patches should be kept as square as practical.

(b) The surface of the concrete around the perimeter of the cut-out areas should be cleaned for a distance of about 100 mm on each side.
(c) The area of the cut-out concrete should be well wetted overnight before the new concrete or mortar is laid. Any standing water should be removed.
(d) The concrete or mortar should be well compacted and then after finishing with a wood float or other method to give the required surface texture, the repaired areas should be covered with polyethylene sheets held down around the edges.

The recommended mix proportions (using volume batching) for the concrete and mortar used in these repairs, are:

Portland cement concrete:

1 part of ordinary or rapid hardening Portland cement
$2\frac{1}{2}$ parts concreting sand (assumed to be damp)
$2\frac{1}{2}$ parts of 10 mm coarse aggregate
The amount of water in the mix should be such as will give a slump in the range of 30–60 mm

Portland cement mortar:

1 part of ordinary or rapid hardening Portland cement
3 parts concreting sand, preferably in the coarser grading

For both concrete and mortar it is advantageous if a styrene butadiene latex emulsion is used as part of the gauging water in the proportion of about 9 litres of emulsion to a 50 kg bag of cement. The use of the latex will increase bond with the base concrete and reduce permeability. However, normal curing with polythene sheets should generally be delayed for 24 hours, or as directed by the suppliers of the latex. If ordinary Portland cement is used, all traffic should be kept off the repaired areas for at least three days, and light foot traffic only allowed for the next four days. These periods can be reduced by about 25% if rapid hardening cement is used. However, if the building is unoccupied and unheated and the weather is cold, these periods should be extended.

When repairs to floors have to be carried out over a weekend so that normal work can be resumed on a Monday morning, it will be necessary to use either an ultrarapid hardening Portland cement such as Swiftcrete, or HAC. Some general information on these two cements is given in Chapter 1. In both cases detailed advice should be obtained from the manufacturers. In the case of HAC, a minimum cement content of 400 kg/m^3 of concrete or mortar and a maximum w/c ratio of 0·40 are essential features of the mix design. Continuous wet curing is also essential.

3.11.2 Repairs to Joints

Floors which carry moving loads from trolleys and fork-lift trucks usually show deterioration at joints before wear appears on the surface within the bays. This deterioration takes the form of fretting and ravelling along the edges of bays and along the wider cracks.

Careful investigation is required before repairs are put in hand as there are many factors involved which will influence the method of repair. These factors include:

The use to which the floor is put, particularly whether the trade is 'wet' or 'dry'
The weight of any moving loads and the type of wheels, *i.e.* nylon, steel, pneumatic etc.
The temperature range within the building

The author's experience is that many joints are not required at all once the initial contraction and shrinkage of the concrete has taken place. The provision of a comparatively wide sealing groove which is filled with a rubber bitumen is often unnecessary. The groove can be reduced to 6–10 mm wide, or in some cases abolished altogether by filling with an epoxide mortar trowelled off smooth, level with the surface of the adjoining bays.

Subject to a decision on whether to 'lock' the joint as mentioned above, or repair it so that movement can take place, the following method of repair will generally prove satisfactory:

(1) The ravelled, broken and otherwise defective concrete along both sides of the joint should be cut away with a concrete saw.
(2) All sealing compound and back-up material in the joint groove should be removed and the joint carefully cleaned out.
(3) The sides of the joint should be remade with an epoxide mortar, leaving a gap of predetermined width (unless it has been decided to lock the joint as described above). Full movement joints are usually about 20 mm wide, while the groove for a contraction joint need not exceed 6–10 mm in width.
(4) After the epoxide mortar laid in (3) above has hardened, the joint groove should be sealed with a preformed neoprene strip which is inserted into the joint under slight pressure with a primer-adhesive.

3.11.3 Repairs to Cracks

In the case of cracks it is even more important to carry out a careful investigation before deciding on what remedial measures to adopt. Cracking in concrete members has been discussed at some length earlier in this chapter, and for the purpose of this section it is assumed that the cracks in the floor slab have no structural significance. The main reasons for repairing cracks of this type in floor slabs are: (1) as a preventive measure to ensure that ravelling and spalling along the edges of the joint does not occur; (2) to make good the edges of the crack where spalling or fretting has already occurred; (3) to prevent penetration of water or other liquids through the slab; (4) when it is intended to lay a bonded topping on the slab all cracks should be properly repaired to avoid their reflection through the new topping.

In determining the method of repair it is necessary to decide whether the crack is 'alive', that is, whether it will continue to move after the repair has been completed. If the answer is in the affirmative then the material used to repair the crack must be sufficiently flexible to accommodate this movement; at the same time the sealant must bond to the concrete. If no further movement is expected then the repair material can be rigid.

Flexible materials include rubber bitumens, specially formulated epoxide resins and polyurethanes, and natural and artificial rubber latexes. Care must be taken in the selection to ensure as far as this is possible, that the material will be durable under the operating conditions of the floor. When the cracks are very narrow, less than about 0·25 mm, the crack filler should be of low viscosity.

For cracks up to about 0·5 mm wide, there is generally no need to cut out the crack. The recommended procedure is to tap lightly along the crack with a chisel, clean out all grit and dust with compressed air, and then brush into the crack a cement latex grout, or a polymer resin.

For wider cracks, particularly when the edges have spalled, the crack should be cut out. A useful tool for this purpose is shown in Fig. 4.22 (p. 166). After cutting out, the crack should be carefully cleaned as described above. If a cement/sand SBR latex mortar is used, it is advisable to wet the crack overnight, and to cure the mortar for four days, starting 24 h after completion. About a month later it will probably be found that a fine hair crack has developed along each side of the repair. This can then be simply grouted-in with a cement/latex grout applied by brush in two coats.

When it is important that the floor is watertight, as may be the case with 'wet' trades, reference should be made to Chapter 4, the section on repairs to floors of water retaining structures.

In concluding this section on repairs to cracks, it should be mentioned that in many cases there may be no need to repair the cracks at all, that is, relatively straight narrow transverse cracks, not exceeding about 0·5 mm wide in a floor slab used for a 'dry' trade with foot traffic or light trolleys.

3.11.4 The Provision of Vapour-Resistant Membranes in Existing Floors

It may be found desirable to insert a new vapour-resistant membrane in an existing floor because either one does not exist or it is clearly defective. Apart from removing the existing floor and laying a new one, it is very difficult to find a satisfactory solution, mainly because change in level between the floor in one room and an adjacent one is usually not acceptable.

If the floor level can be raised, then the procedure is to remove the floor covering, and the screed if it is defective, and clean off and smooth the surface of the base slab. On the prepared surface of the slab, a vapour-resistant membrane should be laid. This can be an insitu material or a sheeting. The insitu materials are usually mastic asphalt, bitumen, tar, or one of the polymer resins such as polyurethane, epoxide, or a combination of both. Sheet materials are polyethylene, butyl rubber, polyisobutylene, PVC etc. It is essential that whatever material is used it should be of adequate thickness and not damaged by subsequent operations.

The insitu materials are applied by brush or spray, or in the case of mastic asphalt, it is spread by hand. The thickness will depend on the material and will vary from 1–10 mm. Sheet materials must be well lapped and all joints should be sealed by hot or solvent welding or by an adhesive which is not moisture sensitive. Welding is better than adhesives. The membrane should be carried up adjoining walls to damp-proof course level.

A new cement/sand screed at least 40 mm thick is then laid on the membrane; the final floor finish is laid as soon as the screed has dried out.

If a change in floor level is not permitted then a rather less satisfactory solution has to be accepted. The existing covering has to be removed, the substrate cleaned off and an insitu membrane laid and the floor covering bedded on this while it is still plastic. This may not prove entirely satisfactory in the long term. It should be remembered that cement-based materials are generally rigid and are not particularly vapour-resistant.

3.11.5 Bonded and Unbonded Toppings

Cases often arise where it is necessary to provide a completely new wearing surface to the whole or major part of a floor area. Such a major repair would probably take the form of a bonded or unbonded topping.

The decision as to which type of topping to use in any particular case will depend on a number of factors:

(a) The condition of the existing floor
(b) The extent to which the floor level can be raised
(c) The proposed use of the floor area after resurfacing

There is detailed discussion of these and other methods of repair in the author's book, Floors—Construction and Finishes, a summary of which will be given here.

Generally, a bonded topping can be used when the base concrete is high quality and is not contaminated with oil or grease. For durability and satisfactory performance, thin bonded toppings rely on full bond with the base concrete as well as high quality in the topping itself. For normal concrete toppings, a minimum thickness of 75 mm is needed for lightly loaded floors and 100 mm for heavily loaded floors.

The following are the main factors involved in providing a satisfactory bonded topping:

Preparation of the base concrete. The concrete should be scabbled with a mechanical scabbler, and all grit and dust removed. The prepared surface should be well wetted the day before the topping is laid.

Mix proportions. The concrete should be batched by weight.

Cement: Ordinary or rapid hardening Portland cement; minimum cement content, 360 kg/m^3 of compacted concrete
Aggregates: These should comply with BS 882: Aggregates from Natural Sources for Concrete; 20 mm maximum size for 100 mm thick toppings and 10 mm for thinner toppings
Workability: A slump of 50–75 mm should be adequate for good compaction
Compressive Strength: With good quality aggregates and the slump mentioned above, a characteristic strength of 30 N/mm^2 should be obtained. The cement content recommended is the minimum. A plasticiser can be used to improve workability when working to a low w/c ratio

It should be noted that abrasion resistance is closely related to concrete strength and this is why emphasis is placed on cement content and low w/c ratio. If special abrasion resistance is required, then either the whole floor or certain selected areas can be finished with a metallic sprinkled finish which is applied to the topping while it is still plastic. The metal is usually finely divided iron and is premixed with the cement and an admixture; the weight of metal can vary from 4–12 kg/m^2 according to the operating condition of the floor.

When the floor is suspended and is used by heavy moving loads, this can cause long term problems with maintenance of full bond between the base concrete and the thin topping. As the loads travel across the slab and beams, reversal of stress occurs and this creates horizontal shear at the interface of the base and topping. In composite construction this is taken care of by means of shear connectors (CP 117: Composite Construction in Structural Steel and Concrete). In the few cases where such conditions are likely to arise, the recommendations in CP 117 should be given careful consideration.

Joints in the base slab should be carried through the topping unless these have been locked as mentioned in the previous section on repairs to joints, or they are truly monolithic. It is unusual to provide reinforcement in thin bonded toppings, but it is advisable to limit the bay size so that the area does not exceed about 25 m^2 and the length should not exceed $1\frac{1}{2}$ times the width. The joints between the bays should be the plain butt type. They may open fractionally as the concrete matures, but this fine gap can be simply grouted-in about a week before the floor is put into use.

The topping must be fully compacted by a vibrating beam and on completion of the compaction and finishing process, the bays should be covered with polyethylene sheets, well lapped and held down around the perimeter by boards.

Regarding the finishing, this can be by power float followed by a skip or bull float. It is probable that early surface grinding will give improved resistance to abrasion, but the most durable finish to plain concrete is likely to result from careful hand trowelling with a steel float.

When unbonded toppings are used, the thickness must be increased so as to give a minimum thickness of 100 mm, and for more heavily loaded floors, 150–200 mm may be required. The thickness of the topping should not vary significantly over short distances, and therefore the base slab may have to have some general repairs before the new topping is laid. It is usually advisable for unbonded toppings to be laid on a slip membrane such as 500 gauge polyethylene sheets. This should prevent cracks and 'live' joints in the base slab being reflected through the new topping. However, all full movement joints in the base slab should be carried through the topping.

The maximum size aggregate should be 20 mm, but the mix proportions, characteristic strength, compaction, finishing and curing can be as recommended for bonded toppings.

The length of the bays will depend on whether reinforcement is used, and if so its weight and the cross sectional area in the longitudinal direction. Unbonded toppings 100–125 mm thick can be laid without reinforcement up to a maximum length of about 3·0 m. With a thickness of 150–200 mm the bay length can be increased to 5·0 m. For bay lengths in excess of 5·0 m, reinforcement in the form of high tensile fabric, placed 40–50 mm from the top surface, should be used. The object of this reinforcement is to control cracking, and the amount of steel in the longitudinal direction is directly related to the length of the bay. The 'long strip' method of laying the topping can be adopted, in which case the bay length is the distance between the stress relief joints. This distance is generally 15 to 25 m.

3.11.6 Repairs with Precast Concrete Slabs

Another effective method of repair to both small and large areas of a floor is to use precast hydraulically pressed paving slabs laid on a cement/sand mortar bed.

These slabs, which should comply with BS 368: Precast Concrete Flags, are strong and impermeable and consequently possess good resistance to abrasion and to a limited extent, mild chemical attack. Where an appreciable degree of chemical resistance is required, clay tiles and bricks should be used, as described in Chapter 6.

The slabs are 50 or 63 mm thick and the mortar bed is likely to be about 25–40 mm thick. The great advantage with the use of these precast elements is that individual slabs can be easily taken up and replaced. To avoid 'rocking' and breakage by impact, the slabs should be fully bedded, and care must be taken to avoid lipping at the joints. The joints can be fine, and simply grouted in with a cement/latex grout.

3.11.7 Other Methods of Repair

There are numerous proprietary materials on the market for effecting repairs to concrete floors. These are particularly useful when small patches have to be repaired as a matter of urgency. They are usually two-pack materials, consisting of a powder and a gauging liquid. The maker's instructions should always be carefully followed.

For complete toppings, apart from the concrete ones already described, there are polymer resin compounds on the market, as well as the more orthodox mastic asphalt, and the rather newer 'cold mastic' jointless and dustless toppings. These latter are proprietary materials, and consist essentially of bituminous emulsion, Portland cement and a mineral aggregate. They will give excellent service under the proper conditions and when correctly laid. Petroleum, and hydraulic oils will cause deterioration in these toppings.

A good hard wearing, jointless and to some extent chemically resistant floor topping is based on polyester resin, Portland cement and a selected fine aggregate, and is marketed under the name of 'Estercrete'. The material is now sold in small packs and can be used for patching as it hardens very rapidly and the floor can be put back into use after 24 h.

3.12 BIBLIOGRAPHY

Addleson, L., The weathering of concrete surfaces. *The Architect*, May 1975, pp. 41–46.

The Concrete Society, *Symposium on the weathering of concrete*, January 1971. Five papers, published by The Society.

Partridge, J., Weathering-design of buildings. *Concrete*, November 1975, pp. 37–40.

Allen, R. T. L., *The Repair of Concrete Structures.* Cement and Concrete Association, London, 1974, p. 20.

Cement & Concrete Association and Cement Admixtures Association, *Superplasticizing Admixtures for Concrete.* Cement and Concrete Association, London (January, 1976).

Perkins, P. H., *Floors—Construction and Finishes.* Cement and Concrete Association, London, 1974, p. 132.

VARIOUS, *International Symposium on Roofs and Roofing*, Brighton, September 1974, 44 papers.
ANON., *Architects' Journal*, Information Library, Everyday Details, and Technical Studies on Roofing.
BUILDING RESEARCH ESTABLISHMENT, *The Developments in Roofing.* Digest No. 51 (2nd series), 1968, HMSO, London, p. 6.
MINISTRY OF PUBLIC BUILDING AND WORKS, *Condensation in Dwellings*, Part 1—Design Guide, 1970, HMSO, London, p. 52.

Repairs to Concrete Building Structures—Part 2

3.13 STRUCTURAL REPAIRS INCLUDING THE STRENGTHENING OF CONCRETE STRUCTURES

Errors in design and the overloading of part of a structure which result in structural failure and collapse are fortunately very rare. When this does happen, a great deal of publicity (some of which can be very ill-informed), is given to the incident. In most cases, when collapse occurs, many factors are involved.

A natural question at this stage is, what is a failure? This is difficult to answer because it is often a matter of degree, and the importance attached to the defect. A failure can be considered as occurring in a component when that component can no longer be relied upon to fulfil its principal functions. For example, deflexion of a floor which caused a certain amount of cracking in partitions, could quite reasonably be considered as a defect, but not a failure; while excessive deflexion resulting in serious damage to the partitions, ceiling and floor finish would be classified as a failure. An error in design may, fortunately, come to light before or during construction, without any harm or damage being caused.

Structural distress in a new unused building, showing as tensile and/or shear cracks of unacceptable width, bowing of vertical members such as columns and walls, and excessive deflexion in horizontal members, indicates that an acceptable factor of safety has been exceeded. This may be due to an error in the original design or some temporary overload during construction. If the latter, then the effect of the overstress on the strength of the structure must be taken into account. The result may be a small reduction in the design factor of safety, which may be accepted when all relevant factors have been taken into account. The limit state design concepts set out in CP 110: The Structural Use of Concrete which includes partial safety factors, can be helpful when considering such cases.

When a design is being checked it is important to look closely at the basic concepts and not limit the work to an arithmetical check. Details, such as depth of bearing of beams, measures taken to ensure adequate rigidity and structural continuity, particularly when precast concrete members are used, should also be carefully scrutinised. It may be prudent to consider the installation of temporary supports pending the result of a full investigation. This enables the whole problem to be reviewed in a calmer atmosphere. Reference should be made to the Report issued by the Institution of Structural Engineers, entitled 'Criteria for Structural Adequacy in Buildings'.

Generally (but not always), cracks which are parallel to the main reinforcement are not as serious as cracks at right angles. However, cracks which extend right through a member should always be repaired. The use of modern low viscosity resins enables specialist firms to seal such cracks so that very little, if any, reduction in strength occurs. The sealing of cracks by resin injection is discussed in detail in the next section in this chapter. Cracks on their own can be repaired by a variety of methods and with a fairly wide range of materials. One of the best known techniques is crack injection using polymer resins. This is particularly suitable when the cracked member has suffered loss of strength, or when the surface chasing out of the crack is considered undesirable. However it is rather unusual for structural repairs to be limited to crack injection.

3.13.1 Crack Injection Methods

For cracks which are deep or which pass right through the member, crack injection can provide a satisfactory solution. It is often difficult to decide whether crack injection will, in any specific case give the better results, compared with the more orthodox technique of cutting away the concrete and repairing with a cement/sand or epoxide mortar. There are of course, many cases where crack injection is the obvious answer, but it is the borderline jobs which are difficult to decide. It is not possible to lay down hard and fast lines on which decisions can be based, but where the concrete itself is good quality, and there has been little or no corrosion of the reinforcement, so that the only real defect is the presence of the cracks, then crack injection alone can be considered as the more suitable solution.

When rusting and spalling have occurred, then the best method of repair may be removal of the defective concrete, cleaning of reinforcement, resin injection, followed by repair with mortar. In this way, the crack within the thickness of the member is filled with resin (or at least partially filled). A final finish of the whole member with a decorative sealant would complete the job and should leave the structure in very good condition.

The essential feature of the resin injection is to inject a suitably formulated resin into the cracks. Correct formulation of the resin is of vital importance. One of the advantages of present day resins is the scope they provide for variations in formulation to obtain optimum characteristics. The requirements for the resin are likely to change from job to job.

The resins most used are epoxide, polyester and a combination of epoxide and polyurethane. The desirable qualities include low viscosity, ability to bond to damp concrete, suitability for injection in as wide a temperature range as possible, low shrinkage, and finally toughness rather than high strength (this latter means a relatively low modulus of elasticity combined with a high yield point). A low 'E' value is of particular importance when further movement is expected to take place across the cracks. Most cracks in concrete are caused by either tension or shear, and the use of a resin with a high 'E' value could result in the concrete cracking again near and probably parallel to the repaired crack.

The work of crack injection can be considered under the following headings:

Preparation of the cracks
Location of injection points and surface sealing
Injection of the resin
Removal of injection nipples (if used) and plugging the hole
Removal of the sealing strip and any final surface treatment which may be required

Preparation of the Cracks

This should consist of removal of any loose weak material on the surface, followed by cleaning of the crack if this is considered necessary. Generally, cracks which are less than 0·5 mm wide are unlikely to require cleaning out unless they have been fouled by the use to which the structure has been put. Compressed air and solvents can be used for this cleaning work and for removal of water in the crack. The author favours the careful use of compressed air as this will remove dust and fine particles as well as water.

Location of Injection Points and Surface Sealing

The distance apart of the injection points will largely depend on the depth and width of the crack. The object is to have as few injection points as possible consistent with maximum resin penetration and ease of filling of the crack, combined with low operating pressure.

After the preparation of the cracks described above, the crack has to be sealed on its surface and the injection points marked. The sealing can be done by a variety of simple materials. The injection points can be either just holes drilled on the line of the crack or they may be nipples screwed into the concrete. Figure 3.14 shows in diagram form, crack injection procedure for horizontal and vertical members.

Injection of the Resin

Crack injection with resins is a comparatively new technique and should only be entrusted to specialist firms, preferably those which formulate and use their own resins. It is natural in these circumstances that each firm

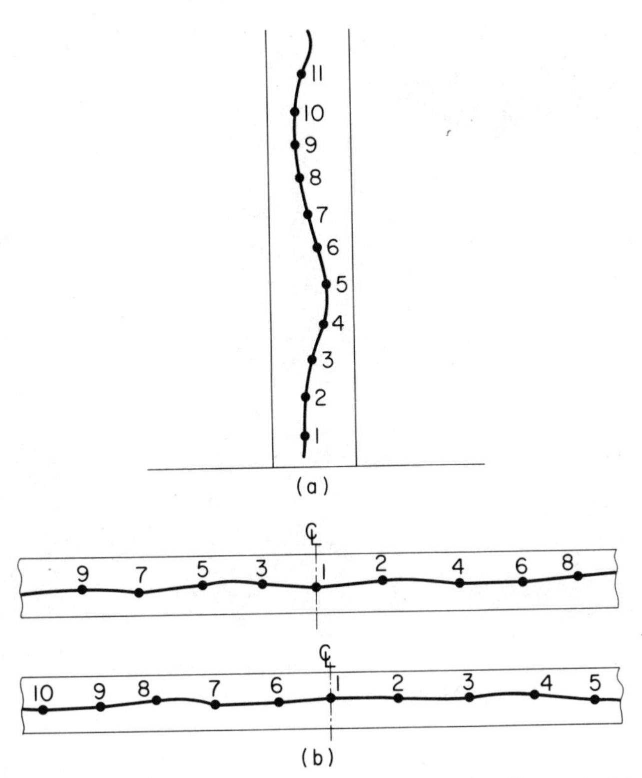

FIG. 3.14. Diagram of crack injection. (a) Injection points for vertical members. (b) Alternative order for injection points in horizontal members.

develops its own technique for the job, and there is considerable divergence of views on the most appropriate method of injection and on the equipment to be used.

Some firms favour simple means of injection by gravity feed or pressure guns, the resin being premixed in batches. Others adopt more sophisticated equipment for continuous feed of freshly mixed resin and hardener through separate feed lines. Figure 3.15 shows a reinforced concrete column being repaired in this way. One firm supplements the pressure feed by the use of a kind of vacuum mat. The author cannot really see the need of a vacuum to aid the penetration, because successful penetration has been achieved on so many occasions without it. As a matter of principle, provided the work is done properly, the simpler the equipment and less complicated the method of application, the better. In appropriate cases, the degree of penetration can be checked by coring, gamma radiography, and ultrasonic pulse velocity survey, but this checking is rather unusual.

FIG. 3.15. Resin injection of R.C. column by the structural bonding process (courtesy: Cementation Chemicals Ltd).

Injection pressures are governed by the width and depth of the crack and the viscosity of the resin. The actual pressure used is low in real terms, and seldom exceeds 1·0 atm (about 0·10 N/mm^2).

What is required with all crack injection, is uniform penetration and complete filling of the crack. A deliberately induced fluctuation in the pressure has been found to be more effective than an increase in continuous pressure. In controlling the injection process it must be remembered that the volume of resin used in filling a crack is very small.

Where the cracks are inclined or vertical, it is usual to commence injection at the lowest injection point and work upwards as the resin emerges from the next higher up injection point.

For horizontal cracks, there is no fixed order of work, as the injection can start from one end and proceed along the crack, or start in the middle and work first left to the end and then right to the end, or alternatively left and right from the middle.

Final Work Following Injection

It is usual to remove the injection nipples (if they are used) and seal the holes as the work proceeds. The removal of the sealing strip can be done after the

FIG. 3.16. Large R.C. manhole cover slab repaired by surface sealing and resin injection (courtesy: Cementation Chemicals Ltd).

resin has cured (which can be from two to seven days), or as soon as it has set which may be within a few hours.

Crack injection, particularly for fine cracks, can be useful when it is important for the repair to be as inconspicuous as possible. Even so, it will be seen. Surface grinding, and some 'cosmetic' treatment will however help to mask the repair.

Figure 3.16 shows cracks in a large reinforced concrete cover into which resin is being injected as a satisfactory alternative to rejecting the unit and making a new one.

3.13.2 STRENGTHENING AND REPAIR

The strengthening and repair of a concrete structure or part of a structure may be required for three principal reasons:

1. When it is intended to change the use of a structure in such a way that the live (superimposed) loads will be increased to an extent that would make a fundamental change in the design necessary. Under the same heading is the case where the use has already changed and the additional loading has caused over-stress in the structural members.

2. Where the structure has for various reasons deteriorated to an extent that its members are no longer able to carry the imposed loads with an adequate factor of safety.
3. Where there is some combination of 1 and 2 above.

The first step is to inspect the structure. With isolated above-ground structures, particularly when they are in an exposed position, it may be found that there is a noticeable difference in the pattern of cracking and extent of the deterioration between the North and South sides. The South side is more exposed to rapid changes in temperature while the North may have suffered more severely from low temperatures and freeze-thaw effects.

While there are still cases of new structures built with poor quality concrete, this is far less frequent than was the case 40 or more years ago. At that time less was known about how to obtain good quality concrete. Concrete was hand-punned and insufficient attention was paid to ensuring an adequate cement content and controlling the w/c ratio, the quality and grading of the aggregates, and the cover to the reinforcement.

The site inspection should include the taking of detailed notes and clear sketches together with high quality photographs which must be properly captioned and dated. The original drawings and design data should be obtained whenever possible. It is usually essential to check as far as it is practicable, the position, condition and amount of the reinforcement. The reinforcement can generally be located with a standard type electro-magnetic cover meter, provided the depth to the steel does not exceed about 90 mm.

In early reinforced concrete structures unusual aggregates were sometimes used. In one case it was found that the aggregate contained ferrous particles which showed on the cover meter as steel reinforcement. Cores taken later showed that in fact there was no reinforcement at all in the slab which was supported on concrete encased steel beams. Some general information on cover meters is given in Appendix 4.

After obtaining information about the existing reinforcement (if any), an assessment has to be made of its structural value. In some cases, due to corrosion it is prudent to ignore it and to provide new reinforcement to take all the design stresses as would be appropriate for a new structure.

Figure 3.17 shows the strengthening of a heavily loaded beam by means of reinforced gunite.

The ideal material for the repair and strengthening of a wide range of structures is high strength reinforced gunite. Gunite is a pneumatically applied material consisting of cement, aggregate and water. There are two guniting processes in general use, but of these only one, the dry-mix process is used for most structural work in the UK. A small amount of wet-mix is used for non-structural repairs and the nozzle velocity is generally lower than that in the dry-mix method.

FIG. 3.17. View of strengthening of heavily loaded floor beam prior to application of high strength gunite (courtesy: Cement Gun Co. Ltd).

The cement and sand is batched and mixed in the usual way and conveyed through a hose pipe by compressed air. A separate line brings water under pressure and the water and cement/aggregate mix is passed through and intimately mixed in a special manifold and then projected at high velocity onto the member being repaired. The force of impact compacts the material, which in good quality work has a density of 2050–2150 kg/m^3 (0·85 to 0·90 that of high grade concrete).

Gunite is used more extensively in the USA, Australia and South Africa than in the UK, although its use here has gained ground in recent years, both for first class repair work and as a structural material in its own right. In the USA gunite is known as 'shotcrete' and there is a Code of Practice for it, namely, 'Recommended Practice for Shotcreting', American Concrete Institute, No. ACI.506-66. In the UK there is no Code specifically for gunite and this has no doubt had an inhibiting effect on its acceptance for new structural elements. For repair work and structural strengthening, it is accepted without hesitation by consulting engineers, local authorities, and government departments.

The cement used in gunite can be either Portland or high alumina; some information on the latter cement is given in Chapter 1. The advantages of correctly made HAC mortar and concrete should not be overlooked where high early (24 h) strength, resistance to dilute acids, and high temperature, are important. It should be remembered that the w/c ratio of high strength gunite is likely to be about 0·35 and this is of great importance when using HAC. The term 'high strength gunite' has been used in this section; for this type of gunite, cores cut from the gunite should have an equivalent 28 day cube strength in the range 60–70 N/mm^2. To obtain such strengths consistently, considerable experience and great care in all aspects of the guniting work is essential.

Before the gunite is applied, the old concrete must be properly prepared, cracks correctly treated and the new reinforcement fixed in position. The preparation of the old concrete consists in the removal of all defective and contaminated concrete. This can be done by percussion tools, grit blasting (wet or dry), or high velocity water jetting. The first two methods are well known, and some detailed information on the use of high velocity water jets is given in Chapter 5.

Reinforcement which is exposed by this cutting away process should be removed if it is severely corroded, or it should be cleaned by wire brushes, needle guns, grit blasting or high velocity water jets.

Cracks wider than about 0·5 mm should be cut out and filled in with either a hand applied cement mortar or with gunite. After this preparatory work has been completed, the new reinforcement is fixed in position. The method of fixing will depend on the type of structure being repaired and the detailed design of the new work.

Some examples of structures which have been repaired and strengthened with high strength reinforced gunite are given in the following sections.

3.13.3 The Repair and Strengthening of Silos, Bins and Hoppers Holding Fine Grained Materials

Silos, bins and bunkers for the storage of granular materials and constructed in reinforced concrete have been in use for more than 40 years. However, precise information on the stresses developed in the walls of these structures during filling and emptying is still lacking.

The American Concrete Institute have produced recommendations for the design and construction of silos, bins and hoppers for holding granular material. The recommendations were prepared by the ACI Committee No. 313 and published in the Journal of the ACI in October 1975. This paper includes a statement that the stresses induced in silo walls by fine grained materials during withdrawal are likely to exceed by many times the stresses caused by the same materials at rest.

A detailed account of an investigation into the cracking in a reinforced concrete cement silo is given in Technical Report No. 296, dated April 1958, by R. E. Rowe of the Cement and Concrete Association, London. The Report concludes that the cause of the cracking was a combination of the effects of the pressure of the fine grained material and the steep temperature gradient through the walls of the silo. The stresses thus developed had not been adequately allowed for in the original design.

An example of a typical repair job follows. A group of reinforced concrete cement silos built about 35 years ago were found to be in a seriously deteriorated condition. Radial cracking had developed at approximately 2·0 m centres and in some places the cracks were up to 40 mm wide. The reinforcement was badly corroded causing considerable spalling of the concrete. The cause of the trouble was diagnosed as errors in the original assessment of the working stresses, poor quality concrete and inadequate cover to the reinforcement. The silos were 17 m high, 5 m diameter, and the walls were 300 mm thick with a number of rectangular openings near ground level. The surface of the concrete was prepared by grit blasting and all cracks wider than about 0·5 mm were cut out and filled in with gunite. After the completion of this preparatory work the new reinforcement was fixed in position on the outside of the walls. This consisted of a fabric (No. A.193—BS 4483) for the complete circumference and full height of the walls; in addition, there were high tensile deformed bars, 16 mm diameter at 200 mm centres for the bottom third of the height, then 12 mm diameter at 200 mm centres for the middle third, and finally, 8 mm diameter at 150 mm centres for the top third. The rectangular openings in the base of the walls required special detailing to ensure that there was no reduction in the strength of the walls at these positions. Each opening was therefore provided with a structural steel frame consisting of two uprights and a head welded together. To these uprights high tensile deformed bars, 16 mm diameter at 200 mm centres, were welded as starter bars; these bars were 500 mm long.

The total thickness of the gunite was 100 mm and the minimum cover to the reinforcement was 25 mm. The total cost of the work was not more than 10% of the estimated cost of demolition and rebuilding.

3.14 REPAIRS TO FIRE DAMAGED BUILDINGS

Fire damage increases each year and in the UK the estimated direct loss in 1965 amounted to about £75 million, while by 1975 it had risen to over £200 million.

Usually after a fire the appearance of the building is such that the owner is in despair and feels that demolition and rebuilding is the only satisfactory course to follow. The total cost of this can be very high indeed, and a major portion of this is the financial loss due to cessation of business from the time of the fire until the premises can be re-opened; this can be anything from six months to two years or more. It has been estimated that on the average, the cost of repairs to the structure is about 50 % higher than cost of replacement of stock. It is therefore important that a realistic assessment of the damage should be made as soon as possible after the fire. When the debris, excess water etc. have been cleared up, a careful examination of the structure will often show that despite the depressing appearance of the building, due to charred finishings, half-burnt stock and smoke blackened structure, most of the building damage is repairable and very little, if any, demolition is needed.

Concrete structures certainly suffer damage when there is a severe fire but this is appreciably less than structures of other building materials. It is sometimes stated that there are many uncertainties about the behaviour of concrete subject to fire. This is true as far as it goes, because the effect of fire on concrete depends on the temperature reached and the length of time that temperature is maintained, as well as the characteristics of the concrete in terms of cement type, w/c ratio, cement content, aggregate type, and the thickness of the concrete cover to the steel reinforcement. In the case of lightly reinforced concrete panels, the thickness of the panel and the temperature gradient through it are also important.

In recent years there have been some ill-informed statements that when polyvinylchloride (PVC) is burnt, chlorine is liberated in such quantities that when it is dissolved in the water from the sprinklers and/or fire hoses, hydrochloric acid is formed in sufficient concentration to attack and seriously weaken any Portland cement concrete with which it may come in contact. Under normal site conditions, a very weak solution of hydrochloric and hypochlorous acids may be formed, but it is most unlikely that the concentration will be at a level which would result in anything more than a slight etching of the surface of the concrete. In the majority of cases even this would not occur.

There may be attack on exposed ferrous metals, but iron rusts in a humid atmosphere in any case unless it is protected. The normal concrete cover to reinforcement would give adequate protection against this form of attack.

3.14.1 Assessment of Damage

In this section the effect of fire on a building structure is being considered and such fires are of comparatively short duration, generally a matter of hours. One of the most important facts to be established as accurately as possible, is the maximum temperature reached during the fire in various parts of the building. This applies to the exposed surface of the structural elements, namely, floors, beams and columns. In the case of floor slabs and beams it is the soffit which is normally the worst affected.

A considerable amount of testing of materials and structural elements is carried out by the Fire Research Station at Boreham Wood, Hertfordshire. The results of these tests are used for the fire grading of materials and components; these gradings are included in the Building Regulations, and expressed in terms of hours of fire resistance. There is considerable argument about the true relevance of the fire test results and the actual effect of a fire in a building on similar materials and components. There are bound to be differences between the artificial conditions surrounding a laboratory test and the conditions in a building which catches on fire. Not only is the temperature different, but the concentration of flame, duration, effect of the presence of other materials, and operation of fire extinguishing equipment, all influence the effect of the fire on the building materials and the structure.

In addition to the change in strength of the materials due to the temperature reached, there will also be the stresses set up in the structural elements by the rise in temperature to the maximum reached during the fire and the return to ambient when the fire is extinguished. The thermal expansion and contraction stresses will depend on a number of factors including the restraint imposed on the elements by other members in the structure and the thermal gradient through the members. These latter are what may be termed 'site effects', and cannot be effectively simulated in laboratory tests.

However, as with most control testing, there is really no practical alternative, and laboratory testing does provide very useful information. Its limitations must however be realised. Experience has shown that in general, materials and structural elements which give good results in fire tests, behave well in actual fires, and *vice versa.*

It is important to ascertain as accurately as possible, the maximum temperatures reached in different parts of a structure during a fire and the temperature gradient through the structural members.

Maximum temperatures can be estimated reasonably accurately by a careful examination of the debris after the fire. It is therefore obvious that the sooner this examination is carried out the better, otherwise it may be found that cleaning-up operations have removed or destroyed valuable information. The temperatures determined in this way are likely to be appreciably different to the temperatures reached by many of the structural elements in the building.

The temperatures reached by the structural members can usually be estimated within certain limits, from a study of the visible changes which occur in concrete, and this information can then be used to decide what additional investigations and testing are required in order to determine the repairs needed.

It is pointed out in the following section that these visual effects may be misleading and should not be considered as conclusive evidence on their own. For this reason, an examination of the debris of other materials will be useful in building up a realistic assessment of the effect of the fire on the structure.

The 'other materials' mentioned above would normally be metal articles, ceramics (clayware), and glass. The time for which the materials in question have been exposed to the temperature is of course important and therefore information on the duration of the fire and its apparent intensity is most relevant.

Some examples are given below, but more detailed information can be obtained from National Building Studies, Technical Paper No. 4, Investigations on Building Fires, 1950, issued by the Building Research Station.

Aluminium and its alloys will form drops at about 600 °C–700 °C.
Cast iron will form drops at about 1100 °C–1200 °C.
Brass will form drops at about 900 °C–1000 °C.
Glass softens at about 700 °C and melts (flows) at around 850 °C.

The fundamental question is what the effect of fire is on a reinforced concrete structure. In providing an answer to this question one has to consider the effect on the constituents of concrete as well as on the steel reinforcement, and on the structure as a whole.

3.14.2 The Effect of High Temperature on Portland Cement

Clearly the effect of high temperature will depend on the temperature reached and the length of time that temperature was maintained. Consideration will first be given to the Portland cement paste, because if this is seriously weakened the concrete is likely to disintegrate.

The uncombined (surplus) water in the concrete will be driven off from the surface layers and some shrinkage cracking in those layers will occur. Up to a temperature of about 100 °C there will be no significant loss of chemically combined water even with a prolonged exposure to this temperature. As the temperature rises above 100 °C there is a gradual loss of the chemically combined water from the calcium silicate hydrates. The actual loss depends on time and temperature. With this loss of chemically combined water there is a drop in the strength of the concrete corresponding to the amount of water lost. However, once the concrete cools down there will be no further reduction in strength.

If the temperature of the concrete reaches 400 °C and above, the calcium silicates commence to decompose into quicklime and silica. This process is irreversible and there is a progressive loss of strength with time. When the concrete cools, the quicklime (calcium oxide) will absorb moisture, converting to slaked lime (calcium hydroxide). When this happens, disintegration of the affected concrete will occur.

3.14.3 The Effect of High Temperature on High Alumina Cement

With temperatures above 100 °C, there is a progressive loss of chemically combined water from the calcium aluminate hydrates. At about 400 °C decomposition sets in but this produces mostly calcium aluminate and alumina which are much more stable than the calcium oxide produced by decomposition of Portland cement. This is basically why HAC is a refractory cement, and with special aggregates, makes refractory concrete.

3.14.4 The Effect of High Temperature on Aggregates

The effect of temperature will depend on the mineralogical classification of the aggregate. For the purpose of this discussion, the aggregates used for structural concrete can be divided into natural and artificial aggregates. The former group is subdivided into igneous, siliceous, and calcareous aggregates.

Igneous rocks, such as granite, basalt etc. are likely to be reasonably stable up to about 1000 °C, although there may be some spalling with a rapid rise and fall in temperature.

Siliceous aggregates, such as flint gravels undergo rapid changes in volume at certain temperatures and this sudden increase in volume may cause disruption of the concrete. The critical temperatures are about 250 °C and 575 °C.

Calcareous aggregates undergo fairly uniform expansion with increase in temperature. At about 400 °C they tend to change to form quicklime (as is the case with Portland cement). On cooling, calcium hydroxide is formed with the absorption of moisture with a serious reduction in strength.

Artificial lightweight aggregates, such as expanded clay and shale, sintered pulverised fuel ash, and blastfurnace slag, are manufactured at temperatures above 1000 °C and are therefore very stable at temperatures below this level.

3.14.5 The Effect of High Temperature on Steel Reinforcement

The effect of temperatures up to about 700 °C on the final strength and ductility of mild steel and hot-rolled high yield steel is from a practical point of view negligible. This refers to the strength and ductility after return to ambient temperature. The effect of the elevated temperature on the reinforcement under load and the disruptive effects of expansion must be given careful consideration.

3.14.6 Investigations and Tests Prior to Repair

In the following paragraphs recommendations are made for the overall investigation of a concrete building structure which has been damaged by fire. The taking and testing of cores and samples of steel reinforcement is suggested in order that a realistic appraisal be made of the condition of the structure and clear recommendations made for repair. The sampling and testing should be done with discretion as it is very expensive, and a great deal of very useful information can be obtained by careful visual examination alone.

After consideration of the likely effect of high temperatures on Portland cement, aggregates and steel reinforcement, this information must be given practical effect by its application to reinforced concrete.

The exposed surface will obviously receive the full heat of the fire and the face temperature will reduce sharply as the distance from the face increases. With temperatures above 400 °C permanent damage is done to the cement paste and therefore this can be considered as applying to the concrete as well. The concrete cover to reinforcement is reasonably effective as thermal insulation and it has been found that in what may be termed 'normal' fires, the temperature of the concrete at a depth of 50 mm from the exposed face, usually does not exceed about 300 °C. In very severe fires, this temperature may be reached at a depth of 75–100 mm. From what has been said, it can be seen that the critical temperature for the concrete is about 400 °C, when the calcium silicates in the cement start to be converted into calcium oxide. Many experienced engineers take this critical temperature rather lower, namely at 300 °C.

It is perhaps fortunate that at this 'critical' temperature, whether it be taken as 300 °C or 400 °C, there is a noticeable colour change, consisting of a pink or pale red coloration. Provided there is adequate lighting, which should in fact always be provided for examination purposes, this pink colour can be easily seen. However, the pink colour may disappear if the temperature reached was very high, but in such a case, the concrete would be very friable. It has also been reported that this colour change sometimes disappears with the passage of time. It is therefore important for investigations to start as soon after the fire as possible.

It should not be assumed that final decisions on the extent of the damage can be taken from visual inspections alone. The visual examination must be supplemented by cutting back the concrete with hand or pneumatic tools. This quickly discloses concrete which has suffered a reduction in strength. In addition, some cores should be taken, and the newly developed ultrasonic pulse velocity (upv) technique has proved very useful.

With large fires, it is always found that some of the reinforcement in the slabs has buckled and this means it must be replaced. Samples can be taken from the buckled bars and tested for yield point and ductility. It is usually found that damage to the reinforcement in slabs is much greater than in beams and this is no doubt due to the greater depth of cover in beams compared with slabs.

It is essential that a carefully thought out system of recording and reporting on the damage to all parts of the structure should be adopted.

This should classify the damage into a number of categories such as insignificant, medium and severe. Each category must be clearly defined by reference to basic criteria such as surface appearance (spalling, colour, cracking, crazing etc.), estimated temperature reached, physical characteristics revealed by use of hammer and chisel and condition of reinforcement. To these can be added, as they become available, the results of core tests, tests on the reinforcement, and if appropriate, upv survey. Each serious fire produces its own set of problems and many papers have been written on the effect of fires of varying intensity in a wide range of structures. Some of these papers are listed in the Bibliography at the end of this chapter.

3.14.7 The Use of Gunite for Fire Damage Repairs

While the normal method of repair is with reinforced gunite, each job brings its own on-site problems and difficulties. A fundamental question which arises immediately is whether a concrete structure damaged by fire and repaired with gunite (a pneumatically applied cement/sand mortar), known as 'shotcrete' in the USA and 'Torcrete' in Germany) will have the same structural factor of safety and fire resistance after the repair as it had before.

In the opinion of the author, the answer, in general terms, is in the affirmative, provided the repairs are properly carried out, and the gunite is of the highest quality. It is normal practice for loading tests to be carried out to 'prove the design' and check the quality of the repair. For these tests the test load is usually 25% above the design live load. Regarding the fire resistance of cement/sand mortar, the following tables are taken from Code of Practice CP 114: The Structural Use of Reinforced Concrete in Buildings, and CP 110: The Structural Use of Concrete. These show quite clearly that cement/sand mortar has at least the same fire resistance in hours, as an equal thickness of concrete. Some authorities consider that high quality cement mortar resists fire better than concrete as it is less liable to spalling.

TABLE 3.1 *Fire resistance of reinforced concrete columns*
(from Code of Practice CP 114 Table 29 B, p. 90)

Additional protection	Minimum dimensions of concrete in mm for a period of					
	4 h	3 h	2 h	$1\frac{1}{2}$ h	1 h	$\frac{1}{2}$ h
None	450	400	300	250	200	150
Cement or gypsum plaster 12 mm thick on light mesh reinforcement fixed around the column	425	375	275	225	175	150

This table clearly indicates that the 12 mm of cement plaster applied on a mesh which is fixed to the column, has the same fire resistance as 12 mm of original thickness of the concrete in the column.

TABLE 3.2 *Fire resistance of reinforced concrete beams* (from Code of Practice CP 110: The Structural Use of Concrete, Section 10, Table 54)

Description	Dimensions of concrete in mm to give fire resistance in hours					
	4 h	3 h	2 h	$1\frac{1}{2}$ h	1 h	$\frac{1}{2}$ h
1. Siliceous aggregate concrete:						
(a) Average concrete cover to main reinforcement	65	55	45	35	25	15
(b) Beam width	280	240	180	140	110	80
2. Concrete as in (1) with cement or gypsum plaster 15 mm thick on light mesh reinforcement:						
(a) Average concrete cover to main reinforcement	50	40	30	20	15	15
(b) Beam width	250	210	170	110	85	70

This table shows that in beams, 15 mm of cement/sand rendering, provided light mesh reinforcement is incorporated, is equivalent to 15 mm of concrete cover.

TABLE 3.3 *Fire resistance of concrete columns (all faces exposed)* (from Code of Practice CP 110: The Structural Use of Concrete, Table 59)

Type of construction	Minimum dimensions of concrete in mm to give fire resistance in hours					
	4 h	3 h	2 h	$1\frac{1}{2}$ h	1 h	$\frac{1}{2}$ h
1. Siliceous aggregate concrete:						
(a) Without additional protection	450	400	300	250	200	150
(b) With cement or gypsum plaster 15 mm thick on light mesh reinforcement	300	275	225	150	150	150

This table indicates that in columns with four hours' fire resistance, 15 mm of cement/sand rendering is equivalent to 150 mm/2 = 75·0 mm of concrete. For two hours' fire resistance columns with a minimum dimension of 300 mm, 15 mm of cement/sand rendering is equivalent to

75 mm/2 = 37·5 mm of concrete. In both cases, the rendering has to incorporate light mesh reinforcement.

The Building Regulations 1972, Regulation E6 and Schedule 8, contain provisions which correspond with those in CP 110.

The Department of Scientific and Industrial Research (now the Building Research Establishment) issued a number of papers on the repair of damaged buildings, and some of these are listed below:

DSIR Note No. 19
The repair of solid concrete and hollow tile floors damaged by fire, 1945 pp. 12. This describes the repair of a fire damaged floor by cement gun method, it also describes the test loads etc.

DSIR Note No. 18
Reinforced concrete columns damaged by fire, Sept. 1945 pp. 12. Page 10 of this note states 'alternatively the new concrete may be applied with a cement gun'.

3.14.8 An Example of Fire Damage Repairs

The author feels that it would be useful to describe briefly an actual fire damage repair which he visited several times during the course of the work. He is grateful to Mr M. Cuckow, Chief Civil Engineer, Watney Mann Ltd, and the Cement Gun Company, for permission to include this information and for the photographs used to illustrate the text.

The building was a large warehouse, and consisted of a steel frame encased in concrete with an insitu reinforced concrete floor. The columns were constructed on a 25 ft × 30 ft grid, with main beams at 30 ft centres and secondary beams at 12 ft 6 in centres. The thickness of the concrete floor slab varied from 10–14 in and was designed for a live load of 560 lb/ft^2. As is not uncommon in such cases, the appearance of the building after the fire was such that complete demolition was seriously considered. However, on the advice of the Cement and Concrete Association, consideration was given to the possibility of repairs in reinforced gunite.

The Cement Gun Company were contacted by the owners, who recommended that one of the badly damaged bays be repaired to their design and then test loaded to 25 % above the design live load, namely 700 lb/ft^2. The live load remained in position for 24 h and the maximum deflexion at the centre of the bay was recorded; the load was removed and then a further period of 24 h elapsed for recovery. The maximum deflexion at the centre of the bay was 0·054 in and the recovery was 94 %. The deflexion was also measured on the supporting beams and these were 0·077 in and 0·092 in with a recovery of 83 % and 98 % respectively.

In this case the damage was for all practical purposes confined to the floor slabs and supporting beams and no significant damage was done to the concrete surround to the steel stanchions. The total area of concrete to be repaired was about 4000 m^2. No two panels had suffered the same degree of damage which varied from very light to severe. A badly damaged panel is shown in Fig. 3.18. Very light damage amounted only to superficial spalling in a few places on the soffit of the slab, while with severe damage almost all the main reinforcement was rendered useless and had to be replaced.

Before the design of the repair could be carried out the true extent of the damage had to be determined. To speed up the work, consideration was therefore given to combining the removal of all obviously fire damaged concrete and a visual examination of the concrete as this work proceeded, with an ultrasonic pulse velocity survey.

It should be realised that there is no generally accepted and recognised method for calculating in advance the resistance moment of a repaired slab because this depends on the degree of bond developed between the gunite and the existing concrete, the strength of the gunite, and the bond between the gunite and the steel reinforcement.

Pulse velocity measurements were made through each slab at nine points and vertically and horizontally through cores taken at the centre of each bay. This information allowed a 'damage profile' to be drawn up for the whole floor, and from this designs were prepared and decisions taken on the location of loading tests which would be most critical after the repairs had been completed.

The work consisted of the preparation of the existing concrete by the removal of all spalled, loose and fire damaged material, followed by the placement of high strength, heavily reinforced gunite.

The visual assessment of how much concrete to remove was co-ordinated with the results of the pulse velocity survey and the two assessments agreed reasonably well.

As previously stated, the actual damage to the concrete varied greatly from one bay to another and even from one part of a bay to another part of the same bay.

Fig. 3.19 shows damaged concrete being removed, and Fig. 3.20 shows new reinforcement fixed in position prior to guniting.

Three designs were prepared:

1. Very severely damaged bays
   (a) All the existing main reinforcement was cut out and all damaged concrete cut away, sometimes to a depth of 75–90 mm.
   (b) Heavy high tensile mesh was then pinned into the sound concrete, with the main wires parallel to the span.
   (c) New 16 mm diameter bars were then fixed, lapped 600 mm at each end to the existing steel which remained embedded in the concrete casing to the steel beams.
   (d) The high strength gunite was then placed to provide a minimum cover of 20 mm.

FIG. 3.18. View of fire damage to soffit of warehouse floor (courtesy: Cement Gun Co. Ltd).

FIG. 3.19. Multi-head pneumatic scabbler removing damaged concrete from soffit of warehouse floor after severe fire (courtesy: Cement Gun Co. Ltd).

FIG. 3.20. Soffit of warehouse floor severely damaged by fire. View of concrete and new reinforcement prepared for gunite (courtesy: Cement Gun Co. Ltd).

2. Less severely damaged bays
   (a) Only the damaged reinforcing bars were cut out, followed by removal of all fire damaged concrete.
   (b) As in (b), (c) and (d) above.
3. Lightly damaged bays
   (a) The whole of the soffit of the slab was grit blasted to a depth of 6 mm
   (b) High tensile fabric was pinned into the concrete, the weight depending on the assessment of the degree of damage.
   (c) High strength gunite was then applied to give a minimum cover to the steel of 20 mm.

An important factor of the repair work was to ensure as far as practical, the application of a uniform high quality, high strength gunite. For this purpose the gunite was checked in two ways, by cutting cores through the repaired slab not less than 28 days after completion of the gunite, and by 'shooting' a large slab of gunite at the same time as the repair was carried out and then cutting cubes from this slab.

The cores and cubes thus prepared were tested by an independent testing laboratory, in accordance with BS 1881: Methods of Testing Concrete. The crushing strength of the cores was found to be in the range of 55 N/mm^2–75 N/mm^2, and the cubes, when tested at 14 days gave strengths above 65 N/mm^2.

In concluding this very brief description of a repair job which took about 12 months to execute, it must be mentioned that the cutting away of the fire damaged concrete proved to be very difficult, unpleasant, and time consuming. The repair was carried out from a high scaffold close to the underside of the slab. The concrete encased steel beams formed a deep barrier around each bay, thus making lighting, and more important, ventilation, of the working area very difficult. When the cutting work was in progress the noise and dust had to be experienced to be believed.

Consideration was originally given to the possible use of high velocity water jets to cut and remove the defective concrete. Owing to the prescribed working conditions in the warehouse laid down by the owners, it was decided that it would be impractical to adopt this procedure.

3.15 REPAIRS TO CONCRETE IN VERY LOW TEMPERATURES

In a limited number of special cases it may be necessary to carry out repairs to a concrete structure in temperatures well below freezing. The two principal cases in which such repairs may be required are:

(a) Structures in the far North or South, close to or within the Arctic or Antarctic zones, or other areas where very low temperatures persist for many months at a time, and (b) in large commercial cold stores which cannot be put out of use without severe and unacceptable dislocation of the business. Repairs under (a), provided the necessary materials and equipment are available, may be technically less complicated than those under (b).

It must be realised that to carry out repairs under such conditions will be very difficult, and the final result is unlikely to be a complete success from the point of view of long term durability. The techniques and materials at present available can be considered only as providing a short term emergency repair.

The first step is to cut away all defective concrete and clean any exposed reinforcement, all as described in this book for general concrete repairs. If the damaged areas are at all extensive it will probably be necessary to repair them section by section. The old concrete which is in contact with the new concrete or mortar must be slowly thawed out and raised to above 0 °C. The new material should be placed and compacted at as a high a temperature as possible, but should not exceed 40 °C. As soon as finishing operations are complete, thermal insulating blankets should be placed in position over the new work and so as to extend some distance around it, overlapping onto the existing concrete by at least 300 mm. From the point of view of maintaining a satisfactory temperature in the maturing concrete, repairs necessitating the use of formwork are easier to execute. It is of great importance that the temperature should be allowed to fall slowly back to ambient, otherwise damage by thermal shock will occur.

For this type of work the use of HAC can be advantageous, as under normal conditions of temperature it reaches its 'maximum' strength in 24 hours. Advice from the manufacturers should be sought, with special reference to cement content, w/c ratio, placing temperature and curing. If Portland cements are used, then an air-entraining admixture should be incorporated in the concrete or mortar, so as to give an air content of $4\frac{1}{2}\%$ $\pm 1\frac{1}{2}\%$. Limestone aggregate in the concrete (if available) will provide a marginal advantage due to its low coefficient of thermal expansion (and contraction).

3.16 THE REPAIR OF EXTERNAL TILING, MOSAIC AND SLIP BRICKS

During the past 15 years or so it has become increasingly popular to use tiles and mosaic as the external finish to buildings. Some architects try to disguise the fact that a building has a reinforced concrete frame (or steel frame encased in concrete), by the use of 'slip bricks' on the external face of beams, and exposed edges of concrete floor slabs.

3.16.1 External Tiling and Mosaic

The use of tiles and mosaic has the great advantage that the surface is virtually self-cleansing and therefore weather staining is reduced to a minimum. Unfortunately there have been a number of cases where large areas of tiling/mosaic have 'failed' due to loss of bond with the background. These cases receive little or no publicity, and there is probably more of this trouble about than is generally known.

For the purpose of this section, tiles and mosaic are considered as one material and will generally be referred to as tiles only.

Due to inaccuracies in the surface of the concrete, it is usual practice to 'dub-out', apply one or more coats of rendering, and then bed the tiles in either a cement/sand mortar or a proprietary adhesive. The best practice is to design, specify and construct the concrete so that the achieved tolerances on the surface finish which has to be tiled are sufficiently close that the tiles can be fixed direct to the concrete without any dubbing-out or rendering. However, this is very difficult to carry out in practice.

Tiles and mosaic used externally on buildings should be frost proof and this means fully vitrified. The tiles vary in thickness from about 10–15 mm, and the mosaic is about 6 mm. The thickness of the bed depends largely on whether a cement/sand mortar or a cement based adhesive is used. When a cement based adhesive is used the thickness would be about 3 mm. The thicker tiles made in the UK have deep recesses on the back, while the thinner tiles (mostly imported from Germany) have shallower recesses.

Reasons for the failures. As stated above, the 'failure' of external tiling is invariably due to loss of bond at one or more levels in the construction. This loss of bond can be slight, medium or severe. In the latter case tiles can fall from the building.

Figure 3.21 shows in diagram form a section through a tiled concrete member as the work is normally executed. It should be noted that if the concrete surface is very irregular, dubbing-out would be needed in addition to the rendering, thus creating an additional layer in various places.

From the diagram it can be seen that there are four planes or interfaces, at each of which loss of bond can occur. For practical purposes, the range of thickness measured from the face of the concrete to the face of the tiles/mosaic, is likely to be as given below:

Tiles		Mosaic	
Two-coat rendering:	15 mm	One-coat rendering:	10 mm
Mortar bed:	10 mm	Bed (adhesive):	3 mm
Tiles:	10 mm	Mosaic:	6 mm
Total:	35 mm	Total:	19 mm

In the two examples given above, the vertical loading is approximately 42 kg/m^2 for the mosaic and 78 kg/m^2 for the tiling. Loading of this magnitude will result in an appreciable shear stress at the interface between the rendering and the concrete substrate, and also to a lesser degree between the various layers.

As everything generally appears to be in order when the job is finished, what occurs subsequently to cause reduction in bond to an extent which results in dislodgement of the tiles? There are many reasons for this loss or reduction in bond, and it is usually difficult to decide with certainty which ones apply in a particular case, and even more difficult to define their relevant importance. It is invariably a combination of factors which eventually leads to the failure.

The principal causes of bond failure may be summarised as follows:

1. Interface: concrete to rendering
   (a) Inadequate preparation of the concrete; for good bond it is

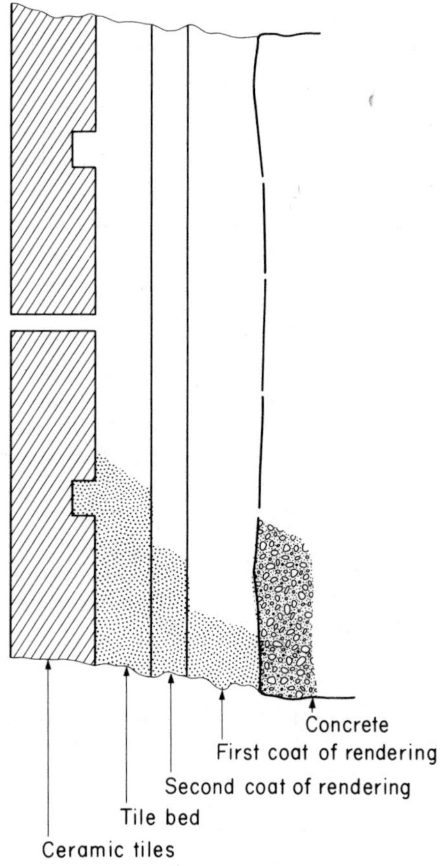

FIG. 3.21. External tiling on reinforced concrete base.

considered at the present time that a mechanical key provided by exposure of the coarse aggregate is more reliable than the use of bonding agents. However, with the further development of bonding agents, this situation may well change. The exposure of the coarse aggregate can be obtained by grit blasting, bush hammering, hand scabbling, or the use of high pressure water jets (the latter is described in Chapter 5). The prepared surface must be clean and free from dust and grit, and preferably slightly damp;

(b) the application of the first coat of rendering to too great a thickness; the maximum recommended is 12 mm;

(c) insufficient compactive effort exerted when applying the rendering;

(d) poor grading of the sand for rendering; also the presence of impurities such as silt, clay and organic matter. The presence of silt and clay particles will show up in the grading, but unfortunately, no specific limits can be set for organic contamination;
(e) inadequate protection and curing of the rendering. The rendering must be protected against strong winds, hot sun, heavy rain and frost, and must not be allowed to dry out too quickly;
(f) corrosion of reinforcement in the concrete resulting in cracking and spalling of the concrete after the completion of the tiling.

2. Interfaces: first to second coat of rendering and rendering to bedding
(a) Lack of an adequate key due to omission of deep combing of the surface of the rendering;
(b) the application of a rich mix onto a leaner mix below. Each coat, including the bedding mortar, should be no richer than the preceding one. The second coat of rendering should also be thinner than the first coat;
(c), (d) and (e) as above.
3. Interface: bedding to tiles
(a) Failure to ensure full bedding of the tiles which includes complete filling of the 'frogs' at the back of the tiles;
(b) the use of too thick a bed;
(c) the use of too rich a mix for the bedding mortar.
4. General—applicable to all layers
Inadequate provision for movement both horizontally and vertically of the various layers including the tiling, when these are considered as a monolithic unit. The concrete which forms the base to the various layers will deform under load, and deformation will occur as deflexion and creep. In addition, the tiles and supporting layers will be subjected to considerable thermal movement, both short term and long term. The concrete structure will be subjected to thermal movements of a different magnitude and a different time scale. The colour of the tiles can have an appreciable effect on the surface temperature reached on hot sunny days, also the orientation, whether North, South, etc. Stress will therefore develop in the various layers due to the thermal gradient from outside to inside. Water penetration can have disruptive effects when it freezes. Horizontal surfaces are much more vulnerable than vertical ones, as the joints between the tiles as they are normally made, cannot be relied upon to be watertight.

Investigations required to assess remedial work needed. While an examination of the original specification and drawings can be useful, it should not be relied upon to give factual information on what was actually done on site. There is no substitute for careful site investigation, and the following procedure is recommended:

(a) Small areas of tiling should be carefully removed, including the bedding and rendering, back to the base concrete. As each layer is removed it should be recorded and preserved in its original state for detailed examination off the site. In addition it may be prudent to carry out a limited amount of coring in those areas which show little or no signs of distress.
(b) The whole surface of the tiling should be carefully examined visually, and by tapping, to locate any cracks, and hollow-sounding (debonded) areas.
(c) The position of any movement joints should be noted as well as the condition of the sealant used, and its type.
(d) The samples of bedding mortar/adhesive and rendering should be of adequate size to enable a thorough examination, supplemented when necessary by analysis, to be carried out. The object of the off-site examination of the samples is to provide additional information so that the reasons for the failure can be established with confidence.
(e) If on removal of the tiles and rendering it is found that the concrete is cracked and/or spalling, then this must also be investigated by the methods described earlier in this chapter.

The investigation outlined above will be expensive as cradles or scaffolding will be required.

Remedial work—methods and techniques. Each job has to be dealt with as a separate problem, because that in fact is what it is. The causes of failure and methods of repair will vary from site to site even though the basic reasons for loss of bond and the principles of repair remain substantially the same.

The most difficult problem which has to be solved realistically relates to the extent of the remedial work, that is how much tiling should be completely removed and how much can remain. The most simple, but most expensive, is to remove the whole of the tiling which exhibits any sign of loss of bond, right back to the base concrete, and start again from there. Recommendations for this new work are not given here as they are described in detail in the latest revision of British Standard Code of Practice CP 212: Part 2: External Wall Tiling and Mosaics, and BS 5262: External Rendered Finishes.

Recommended methods of repair, involving only a limited amount of removal, and preserving the existing tiling as far as possible, are described below.

(1) All tiles, rendering etc., which exhibits significant loss of bond and bulging should be removed back to a sound, well bonded substrate,

which may require going back to the base concrete. There is no reason to remove large areas of sound well bonded rendering where the debonding has occurred only between the rendering and the tile bedding. Considerable experience is required to decide correctly what has to be removed and what can be safely left. While the final decisions may have to be taken on site for some areas, the original investigation and testing outlined above should form the basis for such decisions. The use of percussion tools should be avoided as these set up considerable vibration which can disturb otherwise sound areas. Cutting etc. can be done by saws and in certain cases with high velocity water jets.

(2) The surface of the sound substrate must be provided with a good mechanical key for subsequent layers. For rendering the grading of the sand is most important; a clean sharp silica sand with grading in zone 2 of Table 2 of BS 1199: Building Sands from Natural Sources, is recommended. The surface of the substrate must be clean, and free from dust and loose grit. For the bedding mortar, the recommendations of the suppliers of the tiles should be followed.

(3) Each layer of rendering must be allowed to mature for at least three days before the next one is applied. In cold weather this may have to be increased. During this period the rendering must be protected against strong winds, hot sun, heavy rain and frost. Rendering, because of its very high surface to volume ratio is particularly vulnerable to the effects of low temperature. Deep combing is required to provide a key for the next coat.

(4) Before a new layer is applied, the previous one should be carefully checked for any hollow areas. If the hollow areas are extensive, then the whole section should be carefully removed. Generally, a few small areas can be accepted if they are grouted in with a low viscosity resin as described below. Where the thickness of rendering exceeds a total of 20 mm, a stainless steel mesh should be used, pinned into the concrete with stainless steel pins. Care must be taken to ensure that dissimilar metals are not in contact, otherwise one metal may corrode at the expense of the other. The bedding of tiles/mosaic should not be commenced until the final coat of rendering has matured for at least 14 days.

(5) The most difficult part of the repair is the rebonding insitu of the areas of tiling where some loss of bond has occurred, but not sufficient to justify removal. This assumes that the base concrete is completely sound. These partly debonded areas may be extensive, and the loss of bond may exist at more than one interface in different parts of the area. It is essential to establish at which level the loss of bond has occurred. The usual methods used are resin injection into the hollow areas, and resin anchors. The latter can be

likened to a kind of stitching, or to rock bolts in tunnel construction. A correctly formulated resin for injection must be used, and the required properties are: low viscosity; low shrinkage; low modulus of elasticity; ability to bond to damp concrete and mortar; a reasonable 'pot life' combined with fairly rapid curing; characteristics which will allow its use in as wide a temperature range as possible. If water is trapped between the layers this will interfere with the injection of the resin and steps must be taken to remove it by drilling weep holes and possibly low pressure compressed air. The latter must be used with great care. It is usual to start injecting at the bottom of vertical and sloping surfaces and to work upwards. The principles of crack injection into concrete, previously described in this chapter, apply. The author feels that gravity feed to the injection points is preferable to even low pressure air. Because of the very real practical difficulty of achieving complete filling of all voids, resin injection should be supplemented with mechanical fixing. This consists of stainless steel or non-ferrous anchors. These are proprietary articles. One patented system consists essentially of a glass tube containing the resin which is placed into a predrilled hole. A stainless steel bolt is then screwed in which smashes the glass and releases the resin. The resin sets within minutes of coming into contact with the air and this secures the bolt firmly in the hole. The spacing of the anchors will depend on site conditions but is likely to be about 450 mm–600 mm in both directions.

(6) In view of the fact that the measures described in (5) above are almost invisible when the work is completed, there may be some lack of confidence in the method on the part of the building owner. Therefore a visible technique, in addition to the resin injection, may be more attractive to the client. The method described below was recommended by the author for use on a building overseas, and which appeared to give satisfactory results to all concerned. The technique is to use bands of anodised aluminium (stainless steel can also be used). These bands can be wide or narrow and can run vertically or horizontally or both, depending largely on the degree of support required. The bands are conspicuous, particularly if the background is dark (as in fact was the case). They were considered as an architectural feature and accepted as such by both client and architect. At external angles and the lower edge of projecting beams and slabs, the bands were angles. The bands were fixed to the surface of the mosaic with stainless steel bolts carried through into the base concrete.

Joints in tiling and mosaic. The need for the provision of movement joints in the tiling and the reasons for this, were mentioned briefly earlier in this section. An essential part of any major repair work to tiling must therefore include a thorough review of all movement joints. In some cases it will be found that no such joints have been provided, and this fact will almost certainly be a contributory cause of the failure. It is advisable for these joints to be carried back to the base concrete; a maximum width of 10 mm is all that is likely to be needed. The principles laid down for the materials for joints (back-up materials or inert fillers, and sealants) in Chapter 1, should be followed.

Acceptance Tests

Some architects are enthusiastic about pull-off tests to determine the strength of the bond between the various layers. Sometimes this test is suggested to help decide which areas of tiling need removal and which can be considered as satisfactory. As far as the author is aware there are no figures available based on research, nor on a reasonable number of controlled site tests, from which a minimum pull-off stress can be selected. Experience shows that these tests give a wide scatter of results on toppings which are for all practical purposes well bonded. Bond will never be uniform, and well bonded areas will give support to adjacent areas of lower bond. There are many factors involved in these tests which cannot at present be properly quantified, and the author feels that it is better not to include specific minimum pull-off stress in the specification.

3.16.2 Slip Bricks

Slip bricks, or brick slips as they are sometimes called, are relatively thin pieces of clay brick which are used to mask the outer vertical surface of external r.c. beams and edges of floor slabs. In practice it has proved difficult to ensure satisfactory and permanent bond between the various layers which make up the whole thickness from the concrete to the brick slips. In some cases the brick slips are bedded direct to the concrete, while in others one or more layers of rendering have to be applied to bring the face of the bricks to the correct line.

In the previous section on External Tiling, there is a detailed discussion on bond between the various layers, and all that was said there applies here. However, tiles are usually, but not always, fixed in large areas, while slip bricks are generally in long narrow bands.

There are a number of well informed documents on the problems of slip bricks, details of which are given in the Bibliography at the end of this chapter. The Brick Development Association's Technical Note No. 4, places considerable emphasis on the shrinkage and creep characteristics of the r.c. structural frame, but deals very briefly with the known long term expansion of clay bricks. Nevertheless, the BDA's recommendations for securing the slip bricks, allowing for compression joints at the base of the panel, and provision of a flashing above, are all valid.

The Greater London Council, in their Bulletins (which they emphasise are preliminary expressions of opinion), give very useful and practical guidance on materials for fixing the slip bricks. It appears that their experience indicates that the use of epoxide resin mortar and adhesives, polyester resin adhesives, and cement/sand mortar containing a styrene butadiene based latex emulsion can be satisfactory. The work and tests carried out so far by the GLC show that good bond can be obtained between the slip bricks and the concrete using the materials mentioned above. This suggests that a permanent physical support in the form of a metal bracket below the slip bricks may not be necessary.

Methods for Fixing Slip Bricks to Reinforced Concrete

The following are the essential features of various systems which can be adopted with reasonable confidence. There are no doubt other, more complicated methods, but the author feels that the simpler the method the better.

1. In this method each brick is tied back to the concrete with a stainless steel or durable non-ferrous bolt (gunmetal or phosphor-bronze). This would be very expensive and there are very few cases where such extreme precautions would be justified. A flashing along the top of the slip brick panel and the provision of a horizontal and vertical compression joint, as recommended in 2 and 3 below, should be provided. Ordinary bedding mortar can be used and no special preparation of the concrete is needed beyond general cleaning and brushing down.
2. This method is shown in Fig. 3.22. The slip bricks are fixed with bedding mortar direct to the reinforced concrete base (a beam or floor slab), without any intermediate layers of rendering. Provided the concrete can be finished sufficiently accurately for line and level in the slip bricks to be maintained, this is likely to be the most satisfactory system. The surface of the concrete should be prepared by grit blasting, hand scabbling, or by the use of high velocity water jets, so as to provide a light exposure of the coarse aggregate. All grit and dust must be removed prior to the application of the bedding mortar. If a metal support (stainless steel, gunmetal, or phosphor-bronze) is provided, then the bedding mortar can be normal brickwork mortar with the addition of an air-entraining agent to resist the effects of freeze-thaw. Should the designer decide to omit the metal support, then the author recommends the use of either an epoxide resin mortar or a cement/sand mortar with a styrene butadiene latex emulsion as a gauging liquid. The latex emulsion is usually mixed in the proportion of 10 litres of emulsion to 50 kg

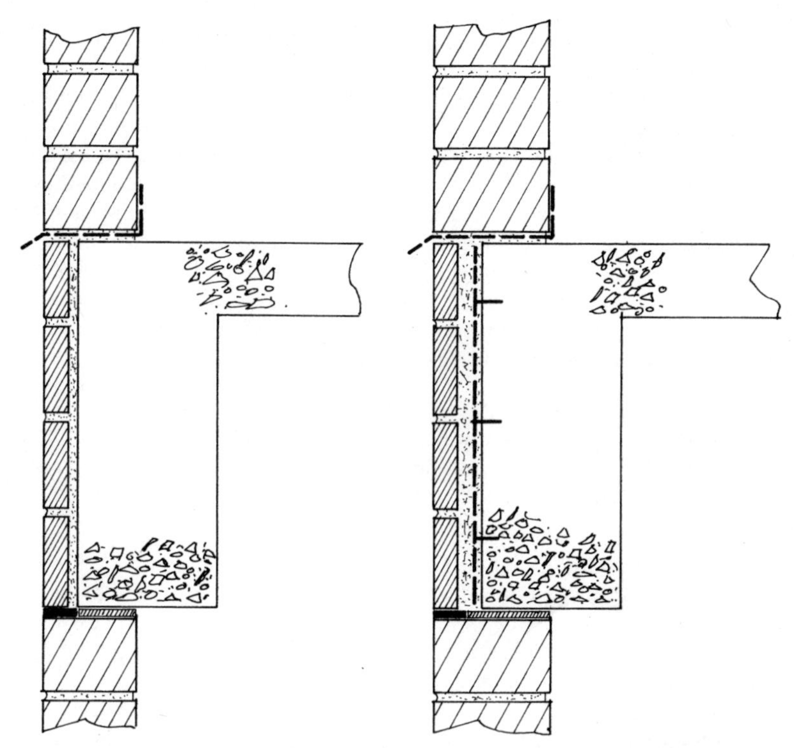

FIG. 3.22. Method of fixing slip brick: bedded directly to R.C. beam.

FIG. 3.23. Method of fixing slip brick: bedded on rendering exceeding 20 mm thick and reinforced with stainless steel mesh.

cement. When the latex is used in a grout as a bonding agent the proportions are generally 25 kg emulsion to 50 kg cement. However, the recommendations of the suppliers should always be followed. The main points to note are the provision of a flashing above the panel carried out beyond the face of the bricks and the compression joint at the bottom of the slip brick panel. This joint can be the same width as the other normal bedding joints in the brickwork, *i.e.* 10 mm; but if a metal support is used this will reduce the thickness of the compression material and some authorities including the Brick Development Association, consider this reduction undesirable, which means a wider joint. In addition to the horizontal compression joint, a similar vertical joint should be provided at each end of the slip brick panel. Sometimes these panels extend right along the face of a long building and the author feels it is prudent to provide intermediate compression joints, at say 15 m centres; however, the need for such intermediate joints is open to argument. If an epoxide bedding mortar is used, then this should be applied in layers not exceeding 5 mm thick as this type of mortar is generally appreciably less cohesive than a cement/sand mortar. The surface of the concrete should be prepared as described above, as a good mechanical key is important. It should be applied direct to the concrete and not to rendering.

3. This method assumes that one or more layers of rendering are required to the concrete in order to bring the surface out to the correct line and level. If the total thickness of the rendering, including any 'dubbing-out', exceeds about 20 mm, then it is advisable to fix a stainless steel mesh as shown in Fig. 3.23. The rendering should be carried out in accordance with British Standard BS 5262: External Rendered Finishes. The use of an SBR latex emulsion in the mix for the rendering is recommended.

BIBLIOGRAPHY

FREEMAN, I. L., *Failure patterns and implications*, Paper given at Joint Building Research Establishment and Institute of Building Seminar, Nov. 1974, p. 13

BATE, S. C. C., *Structural failures*, Paper given at Joint Building Research Establishment and Institute of Building Seminar, Nov. 1974, p. 11

ELDRIDGE, H. J., *Diagnosing building failures*, Paper given at Joint Building Research Establishment and Institute of Building Seminar, Nov. 1974, p. 13

INSTITUTION OF STRUCTURAL ENGINEERS, *Criteria for structural adequacy in buildings*, March 1976, p. 35

LORMAN, W. R., *The engineering properties of shotcrete*, American Concrete Institute. Publication SP14A, Dec. 1968, p. 58

CATON CROZIER, A., Strengthening on fifty year old viaduct, *Concrete*, 1974, July. Reprint, p. 5

LETMAN, A. J. AND HEWLETT, P. C., Concrete cracks—a statement and remedy, *Concrete*, 1974, Jan. Reprint, p. 5

HEWLETT, P. C. AND WILLS, A. J., *A fundamental look at structural repair by injection using synthetic resins*, Symposium on Resins and Concrete, The Plastics Institute and Institution of Civil Engineers, Newcastle on Tyne, April, 1973, Paper No. 17, p. 12

AMERICAN CONCRETE INSTITUTE, Commentary on recommended practice for design and construction of concrete bins, silos, and bunkers for storing granular material. Committee 313, Paper 72-38. *Journal A.C.I.*, October, 1975, pp. 549–565

GREEN, J. K. AND LONG, W. B., Gunite repairs to fire damaged concrete structures, *Concrete*, April, 1971, Reprint, p. 6

GREEN, J. K., Some aids to the assessment of fire damage, *Concrete*, Jan., 1976, pp. 14–17

PARKER, W. T. AND NURSE, R. W., *Investigations on building fires*, Part 1, National Building Studies, Technical Paper No. 4, H.M.S.O., London, 1950, p. 6

MALHOTRA, H. L., The effect of temperature on the compressive strength of concrete. *Mag. Conc. Res.*, **8**(23) (1956), pp. 85–94

DEPARTMENT OF THE ENVIRONMENT/WELSH OFFICE, The Building Regulations 1972—*Guidance Note on Structural Fire Precautions*, H.M.S.O., London, 1975

ANCHOR, R. D., Fire, reinforced concrete. Concrete Practice Sheet No. 20, *Concrete*, Jan. 1975

BRICK DEVELOPMENT ASSOCIATION, Some observations on the design of brickwork cladding to multi-storey reinforced concrete framed structures, *Technical Note* 1971 **1**(4) (Sept.), p. 5

GREATER LONDON COUNCIL, Fixing of brick slips, *Bulletin* (60) (2nd Series), Item 5, Dec. 1972, p. 5

GREATER LONDON COUNCIL, Failure of brickwork cladding to reinforced concrete structures, *Bulletin* (62) (2nd Series), Item 8 (Feb. 1973), p. 3

CONCRETE SOCIETY, Cladding—the provision of compression joints in cladding of a reinforced concrete building. Data Sheet CS1 (Revised) Nov. 1971, Code 52.019, p. 4

ANON., More trouble with tower blocks—bulging brick panels. *Arch. J.*, 1 Nov., 1972, p. 993

CHAPTER 4

Repairs to Concrete Water-Retaining and Water-Excluding Structures

The usual reason for repairing a water-retaining or water-excluding structure is to remedy leakage. To avoid unnecessary repetition, these two types of structure will be referred to throughout this chapter as water-retaining structures and basements respectively.

The leakage may be:

(a) outward from the structure, *e.g.* from reservoirs, swimming pools, sewage tanks etc.;
(b) outward when the structure is full and inward when it is only partially full or empty, *e.g.* reservoirs, etc.;
(c) inward from the surrounding ground, as in the case of a basement.

Associated with any form of leakage, particularly when the structure has been in use for some years, is likely to be the corrosion of the reinforcement and spalling and cracking of the concrete.

It must be realised that in practice, no concrete structure will be what is known as 'bottle tight', unless it is lined with a waterproof membrane. When the structure is properly designed and constructed the amount of leakage should be very small; the amount that can be tolerated will depend on the circumstances of each case. Loss of water from water-containing structures is closely connected with a leakage test, which is normally applied to all new structures. This test is discussed in some detail later in this chapter.

In the case of basements there can be no prescribed water test as such, but basements are expected to be dry. Although in theory, a basement in reinforced concrete can be constructed so as to be completely watertight without resort to tanking, the author's experience is that this desirable aim is unlikely to be achieved in practice. Those responsible for the design of a basement should give very careful consideration to the standard of watertightness required; the design and specification should then be prepared accordingly.

A fundamental principle of repair is to seal the leaks on the water face. Unfortunately this is not always possible, and specialised firms have developed techniques for sealing leaks on the 'wrong' side, that is against the flow of water. When it is proposed to carry out waterproofing work to the inside face of basements, consideration should be given to the possible use of pressure grouting to supplement the surface sealing. The object of the grouting is to help prevent the water penetrating into the body of the wall and corroding the reinforcement.

In the case of old reservoirs, the argument is sometimes advanced that the cost of the water lost by leakage is small compared with the cost of the necessary repairs. This may well be so, but consideration should also be given to the possible effect on the foundations and stability of the structure, by the continuous flow of several thousand gallons per day through the walls and floor.

4.1 THE WATER TEST ON WATER-RETAINING STRUCTURES

Almost all new water-retaining structures, such as reservoirs, swimming pools, sewage tanks etc. are required to pass a water test. Repairs are undertaken when the structure fails to pass the test. In the case of existing structures, when leakage is suspected, it is also usual to carry out a water test. If it fails the test, then investigations are put in hand to determine the location of the leaks, assess why they have occurred, and arrange for their repair. After the repairs are completed, a further test is carried out. Because of this close connection between water test and repairs, the author feels that some discussion on the water test would be useful.

The final draft of the revised Code of Practice, CP 2007 (now BS 5337), states: 'The engineer should specify the permissible drop in the surface level taking into account the losses due to absorption and evaporation. For many purposes the structure may be deemed to be watertight, if the total drop in surface level does not exceed 10 mm in 7 days . . . For open reservoirs tested in a similar manner an additional allowance should be made for loss due to evaporation.' The draft Code recommends that roofs be tested for watertightness when the structure holds potable water. Domed roofs should be provided with a waterproof membrane.

The draft Code recommends that a flat concrete roof should be tested by lagooning the roof slab to a minimum depth of 75 mm for a period of three days. The roof slab can then be regarded as satisfactory if no damp patches occur on the soffit during this period.

It would appear from the wording of the Code that the intention is that the concrete roof slab would be watertight in its own right and that except for domed roofs a membrane is not required. The author considers that the conditions of the test are very severe as in practice it is very difficult to design and construct a flat roof slab so that no damp patches appear.

The author is in agreement with the principle of testing the flat roofs of structures holding drinking water before a waterproof membrane is laid on the roof slab. However, it is prudent to provide a membrane, and the object of the water test would be to detect any leaks which should then be repaired prior to the application of the membrane. Damp patches would under these circumstances not be considered as 'leaks'.

Regarding damp patches, all engineers experienced in reservoir construction know that there is usually condensation on the soffit of the slab. The presence of these patches could cause a dispute to arise. Therefore the slab should be checked for condensation before the roof is ponded; even then, the presence of the water on the roof may result in a drop in temperature of the slab and if it then fell below the dew point, condensation would occur.

While the test is undoubtedly useful, the inherent difficulties mentioned above should be recognised and dealt with in a practical way.

For elevated structures, such as water towers and swimming pools above ground level, where the outside of the walls and the underside of the floor can be inspected, the usual test requirement of maximum drop in level over seven days can be substituted by a clause requiring that there should be no visible sign of leakage.

In considering the Code recommendations, it should be noted that they are not clear. Having said that the engineer should take into account absorption and evaporation, it limits the *total* drop in surface level. In the next paragraph it says that in the case of open reservoirs an additional allowance should be made for evaporation. In the opinion of the author it is better to avoid disputes on site and arrange in the test requirements for the practical measurement of evaporation from both open and closed structures.

Metropolitan Water Board engineers who are among the most experienced in the country, have found that there can be appreciable evaporation in roofed service reservoirs.

A simple and effective method of measuring evaporation is to fill a steel drum with water to within about 50 mm–75 mm of the top and to anchor this in the water of the reservoir during the period of the test. It can be assumed that the drum is completely watertight and therefore any drop in level in the drum will be due to evaporation.

There are a number of theoretical arguments which can be used to show that the evaporation from the drum is different to that from the water surface of the reservoir. No doubt there is some difference, but this is likely to be smaller than the large discrepancy which would result from the use of a formula and the insertion of factors which have to be evaluated by guesswork.

When a water-retaining structure is being filled for the first time (this is usually for the water test), the water level should rise slowly. The author's opinion is that a rate of 0·75 m/24 h should not be exceeded. The majority of structures of this type do not contain more than a 7 m depth of water so that such a structure can be filled in 10 days.

As soon as the structure is full to overflow level, which is usually above operational top water level, the level should be accurately recorded. If a high degree of accuracy is thought necessary, then a hook gauge can be used; otherwise the level can be recorded on the wall or other accessible part of the structure.

The level should be read daily at the same hour; the drop in level should reduce each day so that at the end of the initial 'soakage' period of 7 days, the recorded 24 h drop should not exceed about 2 mm, which is difficult to measure anyway. In certain cases, particularly when the structure has been under construction for a long period, in dry weather, the initial period may have to be extended to 10 or 14 days before relative stability is achieved.

The 7-day test period should then be started by recording the water level in the structure and in the steel barrel used to record the loss due to evaporation. It is advisable to check both levels each day at the same hour.

If the drop in level over the 7-day test period does not exceed the figure in the specification and there are no signs of seepage on any of the exposed surfaces, the structure can be assumed to have passed the test. The question is what should this 'figure in the specification' be. As previously stated, the recommendations in the Code are ambiguous. Many experienced water engineers consider that the loss of water, excluding evaporation, should be appreciably less than 12 mm. It is better to decide on a figure excluding evaporation and insert this clearly in the specification. It is useful to realise the quantity of water lost by a drop in level of say 10 mm; this amounts to 1 m^3 (220 gal)/100 m^2 of water surface. This is an appreciable quantity of water; seepage through fine hair cracks and slightly porous concrete would certainly not result in this amount of loss over a period of 7 days. It is therefore important to include a requirement in the specification that there shall be no visible signs of leakage on any exposed surface of the structure. This will make it clear that all such defects must be put right under the contract without extra payment; also that these slight leakages, amounting to no more than a few gallons a day, will not account for a drop in level of 20 mm or more over the 7-day test period!

4.2 TRACING LEAKS

The next problem is how to trace and locate leaks when the drop in water level exceeds that allowed in the specification. This should present no difficulty with elevated structures, but for those on or in the ground where the underside of the floor cannot be inspected, and sometimes the lower part of the walls as well, considerable difficulty can arise.

The author has recommended in his book 'Swimming Pools', that it is

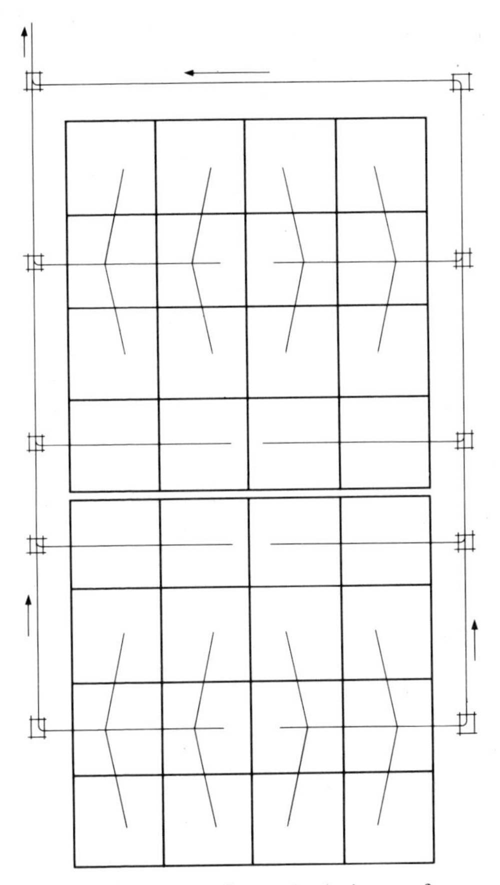

FIG. 4.1. Suggested lay-out for underdrainage of reservoir floor.

worthwhile to spend a little more money than is usual on the underdrainage of the floor. This will enable the underfloor system of drains to be laid in such a way that the floor area is divided into a number of separate units as shown in Figure 4.1. If the external walls are embanked, it is usual practice for a perimeter drain to be laid just below kicker level to drain the embankments. If this perimeter drain is kept separate from the underfloor drainage it will enable leakage through the wall to be seen separately to any flow arising from leakage through the floor. By dividing the underfloor drainage into units, each controlled by a manhole, flow in the system can be monitored, and the search narrowed to a group of about six floor bays.

As previously stated, the most likely places for leakage in a floor are through the joints and these should be carefully examined. Unfortunately, the author's experience is that there can be considerable seepage through joints without any visible sign of defects on the surface.

All this sounds very depressing, and to remedy leaks in floors can be very time consuming and consequently expensive. The Hydrology and Coastal Sedimentation Group of the Atomic Energy Research Station at Harwell, has developed techniques for tracing leaks by means of short-lived radioactive tracers. The Group carries out the work itself and is responsible for the selection of the tracers used in each case. The technique is expensive, but the standard routine of trial and error may cost even more if accurate costing procedures are adopted. The tracer technique is particularly useful where there is no underdrainage, or the drainage system has no access points.

Where underdrainage, as suggested above and shown in Figure 4.1, has been provided, dyes, concentrated salt solution etc. can be used to supplement visual inspections at manholes. However, it must be remembered that dyes will stain the surface of the concrete as well as most other materials. This is important when the structure has already been given a decorative finish, for example a swimming pool.

4.3 REPAIRS WHERE THE LEAKAGE IS OUTWARDS FROM THE STRUCTURE

Water-retaining structures such as reservoirs, swimming pools, sewage tanks etc. are normally designed and constructed in accordance with the British Standard Code of Practice, CP 2007 (revised as BS 5337). One requirement of the Code is that the structures covered by the Code should pass a test for watertightness. When the structure is wholly suspended, as in elevated swimming pools and water towers, any leakage is readily detected. On the other hand, the tracing of leaks in structures built on or in the ground can present serious problems. It is usual for the water test to be carried out before any backfilling is placed around the walls, and this allows any leakage through the walls to be seen. However, if the maximum permitted drop in level in the tank is exceeded, and no leaks can be detected through the walls, then it has to be assumed that the loss of water is through the floor or through the buried pipework, and this is usually very difficult to locate.

Leakage generally arises from three main causes: cracks; porous and/or honeycombed concrete; defective joints.

4.3.1 Repair of Cracks

It has been stated in Chapter 3 that to repair a crack succesfully it is first necessary to know why the crack has formed, and then to decide whether further movement is likely to occur which would cause the crack to reopen if it were repaired with a rigid material.

In new structures, cracking is usually more prevalent in the walls than in the floor and is generally caused by thermal contraction stresses in the early life of the concrete. This type of cracking was discussed in Chapter 3. A typical thermal contraction crack is shown in Figure 4.2. There are differences of opinion among engineers on how these cracks should be

FIG. 4.2. Typical thermal contraction crack in wall.

sealed, with particular reference to whether the sealant should be rigid or flexible.

For roofed structures which are either in the ground or are embanked, the chances of significant movement, once the tank is in operation, are quite small, unless foundation movement occurs. Therefore cracks in this type of structure can be safely sealed with epoxide resins and similar materials which are rigid. It is advisable for the sealing to be carried out as late in the construction process as possible. When the structure is filled with water the cracks will tend to close due to moisture movement as the concrete becomes progressively wetter. In this way, the sealant in the crack, whether it is inserted or injected, will be put into compression. Unless the structure remains empty for a considerable time and the concrete dries out, the cracks should remain closed.

These thermal contraction cracks penetrate right through the wall or floor and therefore constitute a plane of weakness. If they are very fine and have practically or completely sealed themselves while under test by the deposition of calcium carbonate (sometimes known as autogenous healing), they should be repaired as described below. While epoxide resins are normally rigid when cured, specially formulated resins can now be obtained which possess some degree of flexibility. Polyurethanes can be formulated so as to be very flexible. However, for drinking water reservoirs, the sealant should be non-toxic, non-tainting and should not support bacterial growth. Bitumen compounds, if used in large quantities can, under some circumstances, impart a phenol taste to the water.

A decision must be taken as to whether the cracks will be repaired by crack injection methods, as described in Chapter 3, or by a crack filling and sealing technique. A brief description of the latter is as follows:

(a) Carefully tap down the line of the crack with a chisel so as to remove any weak edges caused by honeycombing behind. It should be noted that this is not the same as cutting out the crack. For cracks of this type, the author does not consider that cutting out is either necessary or desirable.
(b) Remove the laitance on the surface of the concrete for a distance of 300 mm on each side of the crack. This can be done by power operated wire brushes, light grit blasting, high velocity water jets, or light bush hammering. All grit and dust must be removed from the prepared surface.
(c) Brush into the crack and onto the prepared area, a minimum of three coats of low viscosity epoxide resin. The resin must be formulated to bond to damp concrete.

Figure 4.3 shows the above work in diagram form.

The author is not in favour of sealing cracks by means of a strip of preformed material such as polyethylene, butyl rubber or polyisobutylene, which is fixed to the concrete by an adhesive. The reason for this is that the adhesive is liable to fail within a year or so because both edges of the strip are exposed to continuous immersion.

Questions are sometimes raised on the subject of 'autogenous healing' of cracks in concrete. The author has not seen an authoritative definition of this term, but it is usually taken to mean that the crack seals itself. This can only occur with very fine cracks, probably less than 0·10 mm wide, provided there is no further movement at the crack. The author's experience is that this 'healing' either occurs within the first week or so or the crack does not really seal itself.

There are cases however, where there is reason to believe that the cracks may open at a later date during the anticipated operating cycle of the structure. This can occur when there is likely to be an appreciable drop in

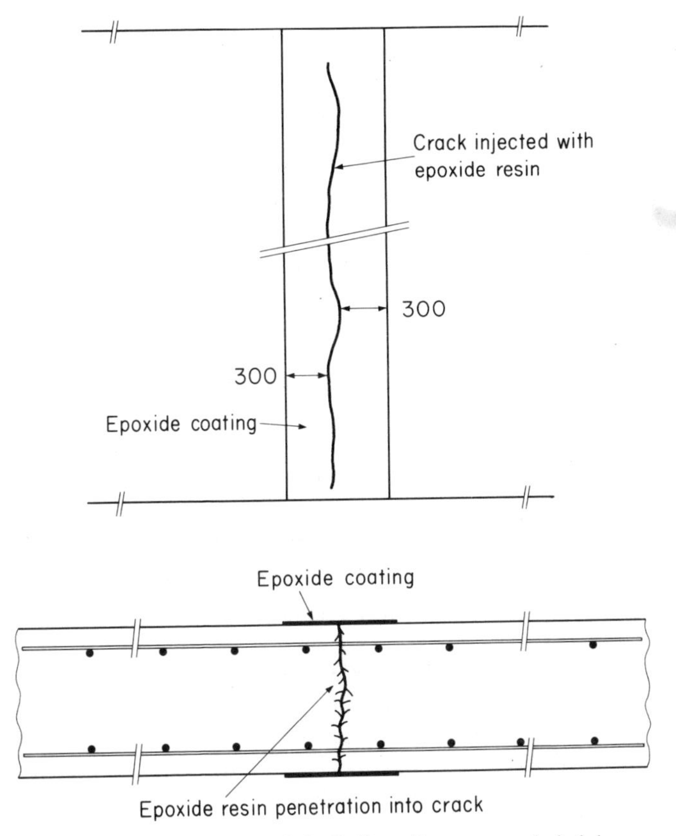

FIG. 4.3. Sealing crack in R.C. wall or suspended slab.

temperature in a structure where provision has not been made for this in the form of movement or partial movement joints. The thermal contraction stresses which develop may cause repaired cracks in the walls or floor to open as they are of course planes of weakness in the structure.

An obvious example of this type of operating cycle occurs in a swimming pool. The water is maintained at a temperature of about 27 °C, and the air temperature in the hall at about 29 °C for a period of three to six or more years. Then the pool is emptied for complete cleaning, maintenance and repair, usually in the middle of winter. This can result in a drop in temperature of 20–25 °C. Pools which are elevated are particularly vulnerable to this, as the area below and around the pool shell is often used for plant rooms and stores etc. When the pool is emptied, the opportunity is taken to service the plant and equipment as well.

The problem is how to repair cracks so as to allow a certain amount of flexibility and at the same time to ensure water-tightness, and this can present some difficulty. If the movement across the crack is expected to be very small, then the use of an epoxide resin specially formulated to provide a degree of flexibility, is probably as good as anything at present available. This would be suitable when there are a number of cracks parallel to each other and it could be assumed that the total movement would be distributed over say two or three cracks instead of concentrated across one. In such a case, *i.e.* where it is decided to use a flexible epoxide resin, the crack should be cut out to a width of 20 mm and to a depth of 20 mm; this groove should then be carefully filled in as directed by the supplier of the resin.

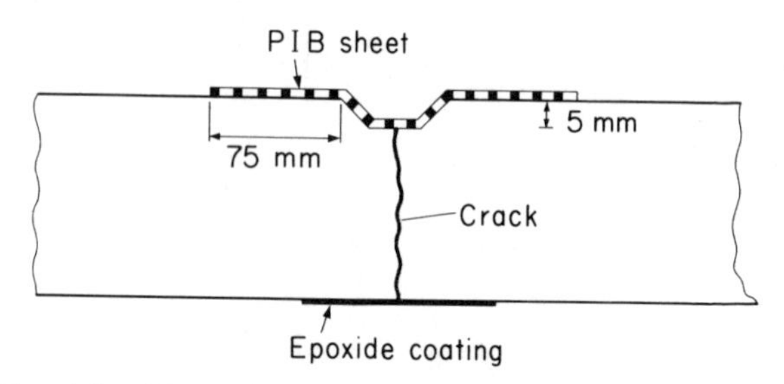

FIG. 4.4. Alternative method of sealing crack in R.C. wall.

Crack injection as described in Chapter 3, using a resin of low viscosity and low 'E' value may also provide an acceptable solution. It must be remembered that the sealing of cracks in the structure itself with a semi-flexible material, does not in any way solve the problem of possible cracking in rigid materials, such as tiling or protective chemically resistant rendering, which may be applied later.

When it is considered that the possible movement across the crack will be greater than can be accommodated by the flexible epoxide resin, then another method should be adopted. The detailing of such repairs requires careful thought and the following is one solution, but this should not be considered as necessarily applicable to all cases.

(a) A channel should be cut in the concrete for the full height of the wall, so that the crack is within the cut-out section as shown in Figure 4.4. The depth of this channel need not exceed 5 mm. The sides of the channel should be cut with a saw, and the surface of the channel should be smoothed with a carborundum wheel. Following this preparatory work all grit and dust must be removed.
(b) A low viscosity epoxide resin should be injected into the crack. The resin should bond to damp concrete and possess some degree of flexibility. Special care is needed for this type of injection. If the anticipated movement across the crack is likely to exceed the safe

extension of the resin it is better not to use resin injection at all. The reason for this is that the resin may bond so strongly to the concrete that when tension develops across the crack, the concrete may fracture nearby, usually parallel to the original crack.

(c) Line the channel with polyisobutylene and carry this along the face of the concrete to a minimum distance of 75 mm each side of the channel, all as shown in Fig. 4.4. The PIB should not be less than 1 mm thick and should be bonded to the concrete with an adhesive which is not moisture-sensitive.

If the walls are to be rendered or given some other applied finish such as ceramic tiles, then the detail would be somewhat different, and a suggestion for this is shown in Figure 4.5. It should be appreciated that the crack in the wall will not be directly below the joint in the rendering and tiling and

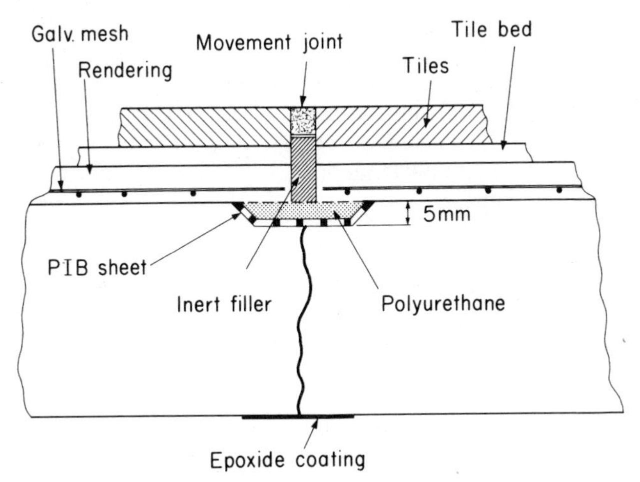

FIG. 4.5. Method of sealing crack in tiled R.C. wall.

therefore complete absence of subsequent cracking in the rendering etc. cannot be guaranteed. Figures 4.4 and 4.5 show resin coating across the crack on the outside face of the wall. Obviously this can only be applied if the outer face is accessible. The author is in favour of the use of a rigid (hard) resin for this external coating because if this fractures, it would indicate that movement across the crack was taking place and alert supervisory staff.

4.3.2 REPAIR OF POROUS AND HONEYCOMBED CONCRETE

The first decision to be taken is whether the whole of the defective concrete should be cut out and replaced with new concrete or whether pressure grouting and/or surface sealing can be relied upon to ensure watertightness and protect the reinforcement. The author knows of no rules which can be used for such decision making but there are certain practical considerations which should be kept in mind.

(a) First, one should ascertain as far as practical, the area and depth of the defective concrete. This can be done conveniently by an ultrasonic pulse velocity survey. The Cement and Concrete Association carry out such surveys on a fee basis. The alternative is coring, which is expensive and time consuming and of course causes some damage to the concrete, and only a comparatively few cores can be taken compared with the large number of upv survey points.
(b) If concrete is cut out it has to be replaced and careful consideration should be given as to how this can be done so as to ensure proper compaction and bond to the old concrete. In vertical members this can be very difficult. Also, for example, if a wall panel is to be demolished and the joints between the panels are provided with water bars, the part of the water bar in the panel being demolished is very likely to be severely damaged. In such a case, how will this be rectified? The answer that it is the contractor's responsibility does nothing to solve a very real practical problem. There is a further discussion on this particular problem in the section dealing with the repair of joints.

Cutting Out

If, after a careful investigation it is decided to cut out the defective concrete, then the method of cutting out and replacement must be considered. An important question is what material to use to replace the concrete which is removed. The various alternatives are:

(a) concrete; either of the same mix as originally used, or a fine concrete using a 10 mm coarse aggregate;
(b) cement/sand mortar, either hand applied or pneumatically applied;
(c) a resin mortar.

(a) *Repair with concrete*. Having decided that the defective concrete must be removed, the first matter to be settled is the most suitable method of removal. If a complete bay has to come out, this is generally simpler than cutting out part of a bay or panel, but as previously stated, the presence of water bars can prove a great embarrassment.

Generally, water-retaining structures are fairly massive, and the vibration set up by the use of pneumatic tools will not cause any damage. If for any reason vibration must be reduced or avoided, then high velocity water jets can be used to cut the concrete into sections; the exposed reinforcement is then cut with oxyacetylene equipment. A further

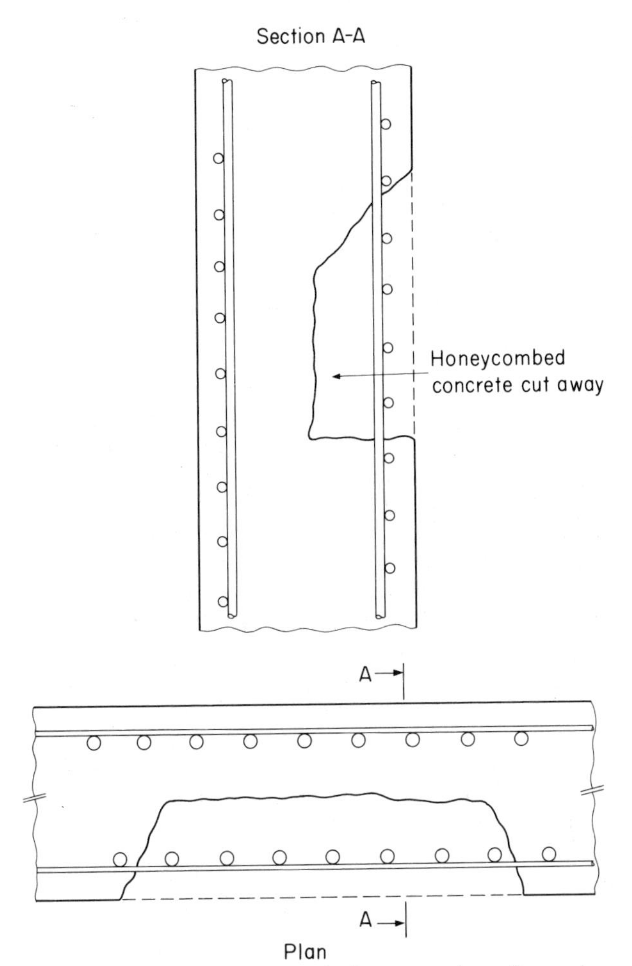

FIG. 4.6. Repair of honeycombed concrete in wall or column.

advantage of high velocity water is that it leaves the surface of the concrete in good condition for obtaining excellent bond with the new concrete without further treatment. When percussion tools are used, the surface of the old concrete must be cleaned of all grit and dust prior to the placing of the new concrete.

For the new concrete it is recommended that the mix be rather richer than that used for the original concrete, and special care should be taken to ensure good workability to aid the difficult job of compaction. Figure 4.6 shows suitable arrangements for a repair to a wall. The need for maximum compaction of the concrete cannot be over-emphasised. Due to the nature of the work this can be very difficult and the use of one of the new 'super plasticisers' may be of great help; some information on these is given in Chapter 1.

Apart from compaction, there is the problem of ensuring a watertight joint between the old and new concrete; this assumes even greater importance if the cut-out section penetrates right through the wall. A clean, dust free surface to the old concrete is essential. While a brush coat of grout on the surface of the old concrete, applied immediately before the new concrete is placed, will help ensure good bond, it is essential that the concrete placing follows immediately behind the grout. If this cannot be ensured then it is better to omit the grout. If a grout is used, it is recommended that it consists of 2 parts ordinary Portland cement to 1 part styrene butadiene latex emulsion by weight. If 10 mm aggregate is used, then the sand content should be increased to about 45 % or even 50 % and there should be an increase in the cement content. The new concrete should be cured in the usual way for at least four days.

As late as possible, before commissioning or testing, the perimeter joint between the new and old work should be sealed. The surface of the new concrete and the old concrete for a width of 150 mm should be wire brushed to remove all weak laitance, and then brushed to remove dust etc. The material used to seal the joint can be either the previously mentioned cement/latex grout (2:1) or an epoxide resin or polyurethane. A minimum of two coats is required, applied by brush, the second coat at right-angles to the first. For drinking water reservoirs, the resin must be non-toxic and non-tainting.

(b) *Repair with cement/sand mortar*. For small areas, particularly those of shallow depth, the use of a cement/sand mortar is often a practical and satisfactory solution. The honeycombed concrete is cut out as previously described in (a) above, and the void is carefully filled with the mortar. The mortar should have as low a w/c ratio as possible so as to reduce drying shrinkage. The actual consistency of the mortar will depend on the exact condition of placing. In some cases, a very dry mix (earth dry) can be used and compacted into the void with hand or pneumatic tools. Honeycombing at a kicker joint at the base of a wall can often be repaired in this way.

For normal hand applied mortar, the mix proportions are usually about 1 part OPC or RHPC to 3 parts well graded sand to Table 2 of BS 1199, Zone 2. The author recommends the use of a styrene butadiene based latex emulsion as a gauging liquid, with water added only to obtain the necessary workability. The use of the latex will improve bond with the base concrete, and help to reduce permeability and shrinkage.

Effective and practical measures should be taken to cure the mortar in the repaired areas, but if an SBR latex emulsion is used, the start of curing is usually delayed by 12–24 h, or as directed by the suppliers. For larger areas, particularly those of relatively shallow depth, say down to 100 mm, pneumatically applied mortar is likely to be a better solution than hand applied mortar. The reason for this is that the mortar will be better compacted and therefore denser and more impermeable. Gunite, in which the mortar leaves the nozzle of the 'gun' at high velocity, is a structural material in its own right. The use of this material has been described in detail in Chapter 3, to which the reader is referred. The usual mix proportions are 1 part OPC to 3 parts well-graded concreting sand, with a w/c of about 0·35. The resulting mortar can have a compressive strength in excess of 50 N/mm^2. It is of course only practical to use gunite where there is sufficient repair work to justify bringing the equipment to the site.

With hand applied mortar, and to a lesser extent with gunite, a fine hair crack is liable to appear around the perimeter of the repaired area. To seal this, it is recommended to brush the new mortar with a stiff bristle brush, and the surrounding concrete with a wire brush, to remove any relatively soft surface layer. This treatment should extend for 150 mm each side of the boundary between the old and new work, and should be carried out as late in the construction process as possible. As soon as this preparation is completed, a heavy brush coat of grout should be applied and worked well into the surface. The grout should be composed of 2 parts OPC and 1 part styrene butadiene latex emulsion by weight.

A new channel section neoprene gasket has come onto the market and is shown in Figure 4.20 (p. 165). If this had been available earlier it would probably have provided in many repair cases a simpler and quicker solution than that described above.

(c) *Repair with epoxide resin mortar*. An epoxide mortar normally consists of the resin, the accelerator and a fine silica sand. It is important that the resin should be formulated to meet the conditions under which it will be applied and be required to operate. General information on epoxide resins has been given in Chapter 1, to which the reader is referred. An essential feature of the resin is that it should be capable of bonding to damp concrete, unless the repair is inside a building and the concrete is known to be dry.

Epoxide mortars are very expensive compared with cement/sand mortars and so their use is only justified in special cases. Also, there will be a greater difference in colour between the epoxide mortar and the surrounding concrete. When they are applied to a vertical surface, the maximum thickness of each layer should not exceed about 5 mm. The reason for this is that the resin mortar is much less cohesive than a cement/sand mortar.

For drinking water reservoirs, the resin should be non-toxic, and non-tainting, and must not support bacterial or fungoid growths.

Surface Sealing of Honeycombed Concrete

In some cases it will be decided that cutting out the concrete is not necessary or not desirable, and surface treatment of the area might then be adopted as an alternative to pressure grouting or in addition to pressure grouting.

The application of a resin coating of suitable formulation is only practical where the surface of the concrete is reasonably sound, and the honeycombing is below the surface. Sometimes interior honeycombing shows itself by a few surface defects and it is only when an ultrasonic pulse velocity survey is carried out or cores are taken, that the true extent of the honeycombing is revealed.

The author feels that the strongest argument against a coating is that sooner or later it will have to be renewed. However, this also applies to protective treatment of ferrous metals and to the sealants used in joints. The great advantage of this method is that all parts of the process (surface preparation and application of the coating) are visible and open to inspection, whereas pressure grouting cannot be seen and is something of a hit or miss business when used on its own. The surface of the concrete should be prepared by removal of the laitance and light exposure of the coarse aggregate. An exposure of 3 mm is the maximum which is needed, and 1 mm is usually sufficient.

It is generally advisable for the resin to be formulated to bond to damp concrete, and of course it must possess adequate durability under the operating conditions of the structure. Epoxide resins and polyurethanes, and sometimes combinations of both, are the resins in general use.

It is advisable to use a low viscosity primer, followed by successive coats of a high build resin to give a total thickness of not less than 0·75 mm. The treated area should extend well beyond the estimated boundary of the honeycombing, *i.e.* by not less than 300 mm.

For drinking water reservoirs all repair materials used for the repair work should be non-toxic, non-tainting and should not support bacterial growth.

Pressure Grouting of Honeycombed Concrete

When a wall or floor is leaking due to porous concrete, a cure can often be effected by pressure grouting. A description of pressure grouting to seal leaks in basement walls and floors is given in Advisory Leaflet No. 52, issued by the Department of the Environment. The impression given by the Advisory Leaflet is that pressure grouting is relatively straight-forward and that if the procedure described is followed, the leaks will be sealed. Unfortunately the author's experience has caused him to be rather pessimistic about the results of pressure grouting on its own unless it is carried out by an experienced firm. Even then, disappointments are not uncommon. It is relevant to note that specialist firms will very seldom guarantee success and insist on working on a time-plus-materials basis. This means that they will not give a fixed price nor will they guarantee complete watertightness on completion.

The grout used is usually based on Portland cement with an admixture,

FIG. 4.7. Pressure grouting concrete backing to brick facing of viaduct (courtesy: Colebrand Ltd).

which is often PFA. Sometimes non-shrink grouts are used. These are often premixed powders to which water is added. Another method is to use a special admixture in the mixing water, which the manufacturers claim will reduce or even eliminate shrinkage.

In the author's opinion, it is best to combine pressure grouting with surface sealing. The surface sealing in existing (not new) structures can usually be applied only to one face of the defective concrete. This technique is discussed in more detail later in this chapter in the section on Basements (4.4).

Figure 4.7 shows pressure grouting of concrete backing to brick facing.

4.3.3 Repair of Defective Joints

The author's experience is that defective joints are the principal cause of leakage in water-retaining and water-excluding structures. Joints of any kind in this type of structure can be termed a necessary evil. It is impossible to build a liquid-retaining structure without joints; in addition, the joints must be correctly located and correctly detailed. It is unfortunate to say the least, that some designing authorities leave matters relating to joints to the contractor and then blame him when leaks occur.

This attitude is certainly contrary to the intention of the relevant clauses in Code of Practice CP 2007 (now BS 5337). The Code makes it clear that the location and detailing of joints is an essential part of the design and is the responsibility of the engineer. Joints detailed and designed to allow thermal movement to take place must be provided unless the engineer is satisfied that other satisfactory provision has been made to control thermal contraction cracking. There are numerous papers on this important subject and a reader who wishes information should refer to the Bibliography at the end of this chapter.

It is normal practice, although some engineers do not agree with this, to provide water bars in joints. In walls and suspended slabs, these are usually the centrally located dumb-bell type. In addition, the water side of the joint is usually provided with a sealing groove which is filled with a sealant. However, for what are known as monolithic (construction or day-work) joints, some engineers rely entirely on the bond between the old and new concrete for watertightness.

Figures 4.8 and 4.9 show defective joints in a reservoir before and during repair.

In spite of these rather elaborate and expensive precautions, joints often leak. In the case of walls and suspended slabs, the location of the leak is clearly visible, but with floor slabs supported on the ground, the tracing of the leak is very difficult indeed. Most structures of this type are underdrained, and if the layout of the drainage system has been designed with this problem in mind (*i.e.* each section of the underdrainage deals with a particular area of floor), then the locating of the leak is much facilitated.

FIG. 4.8. Defective joint in reservoir wall (courtesy: Thames Water Authority).

Even so, there is likely to be an appreciable area of floor which is under suspicion. Once it is established that the leakage is through the floor, then in most cases it is fairly certain that it is the joints which are responsible. The flow through the underdrains must be carefully monitored as each length of joint is ponded. This is a time consuming and difficult job and requires patience and care. Sometimes reference to the daily reports from the R.E. or Clerk of Works will be useful in revealing some forgotten hitch in the work which may have resulted in undercompaction of concrete around a water bar, or displacement of a water bar during concreting. It is not possible in a book to discuss all the possible trouble which can occur at a joint and so a typical example will be given to illustrate what the author considers are important principles.

FIG. 4.9. Defective joints in Fig. 4.8 during repair with 'Neoferma' neoprene gaskets (courtesy: Thames Water Authority and Colebrand Ltd).

Figure 4.10 shows the situation as finally ascertained after lengthy investigations following the failure of a sewage tank to pass the prescribed water test. The displaced water bar was located by careful drilling alongside the joint. There were welded connections at C and D where the displaced water bar CD joined water bars EF and CK, and DJ and GH. T_1 and T_2 are the position of the two drill holes which located the displaced water bar.

The problems involved, and the line of argument developed, may be stated as follows:

1. As the water was passing through the joint, it was obvious that neither the sealant nor the water bar were satisfactory. It was

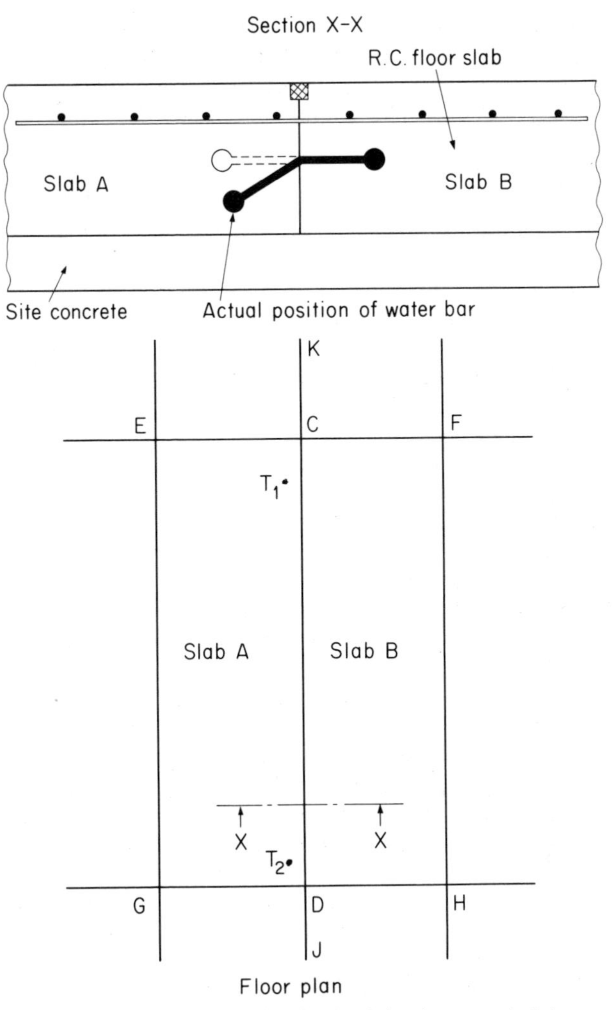

FIG. 4.10. Repair of defective joint in ground slab.

relatively easy to remove the sealant (which was a rubber bitumen) and replace it. This may have cured the leak.

2. However, it was known that sealants are not long lived materials, particularly rubber bitumens and poly-sulphides. Although neoprene has a much longer life, it is also likely to deteriorate in the course of time. As soon as deterioration starts and/or bond with the sides of the sealing groove begins to fail, leakage will recommence because the water bar is in some way defective. The water bar was

specified as an additional safeguard, but due to some failure by the contractor, it no longer fulfilled its purpose.

3. The contract was already behind schedule and the tank was urgently needed. The contractor accepted responsibility for the remedial work.

The real question then was, what was a practical and reasonably satisfactory method of repair?

One suggestion was to cut out a strip of concrete about 1·00 m wide parallel to the joint CD. Another idea was to completely remove bays A and B, but this immediately raised the question of how to deal with the water bars in the joints EG, FH, EF, and GH. A further suggestion was to try to pressure grout the defective concrete around the water bar in bay A. This was rejected as it would probably result in the grouting-in of part of the underfloor drainage.

The author's suggestions are given below; these were based on the principle that as little concrete as possible should be cut out.

(a) The existing sealant should be cut out and the sealing groove deepened by about 15 mm.
(b) Both sides of the sealing groove should be reformed by first cutting the concrete with a saw to widen the joint, and then trueing-up the sides with an epoxide mortar.
(c) After the reformed groove had been carefully cleaned out, a cold-cured epoxide resin should be inserted to a depth of about 15 mm. This resin should possess low viscosity, flexibility and should bond to the damp concrete sides of the groove. The reason for providing this under layer of resin was that the final (top) sealant described in (e) below, would not bond to the existing sealants in the adjacent joints.
(d) After completion of (c), a cellular neoprene jointing strip should be inserted in the groove above the epoxide resin under layer. The strip would be bonded to the concrete sides of the groove by a suitable primer. It would not be bonded to the epoxy under layer.
(e) For a distance of 600 mm each side of the joint and for its full length, prepare the surface of the concrete by mechanical scabbler, grit blasting or high velocity water jets, so as to lightly expose the coarse aggregate to a depth of about 5 mm.
(f) An epoxide resin mortar should be laid on the prepared surface of the concrete to a minimum thickness of 5 mm, the resin to be specially formulated to bond to damp concrete.

The work was carried out generally as described, and no further leakage was observed. It was completed in five working days, compared with an estimated minimum of three weeks for any of the alternative schemes. The formation of additional joints, with corresponding hazards of leakage, was thus avoided.

4.3.4 General Repairs and Lining of Water-Retaining Structures

While the protection of concrete against chemical attack is dealt with in Chapter 6, there are many borderline cases of mild chemical attack on concrete in service reservoirs due to certain aggressive characteristics of the raw or partially treated water. This occurs particularly where the water is derived from an upland gathering ground.

Such waters are characterised by:

Low total dissolved solids (TDS)
Low total hardness
Carbon dioxide in solution (aggressive, or free CO_2)
Low pH
Low alkalinity
Organic and other acids in solution

Occasionally, the acids include sulphuric derived from the breakdown of sulphur compounds in peat and marshy ground by bacteria. The presence of sulphuric or other mineral acids will greatly increase the severity of attack. It is often found that the pH varies over a range from about 4·0–6·5. In some parts of the country there is a significant drop in pH after heavy rainfall.

While it is customary in many cases to treat the raw water before it enters the reservoir by filtration through limestone chippings and/or addition of lime, with sudden variations in the acid content of the water the pH correction may not always be as successful as one would desire.

These facts are mentioned because they have an obvious bearing on the assessment of any damage done and on the method of repair.

Repair of Etched Surfaces

Attack by these soft, rather acid waters normally shows as etching of the surface of the concrete. The attack is in fact the chemical reaction between the acid in solution and the alkali in the cement paste, and on carbonates (if any) in the aggregate. This type of attack is discussed in some detail in Chapter 6 to which the reader is referred.

The etching will vary considerably in depth. It may be only a light exposure of the coarse aggregate, but on the other hand with a rather poor quality concrete and a water with a high free (aggressive) CO_2 content, the attack can be severe.

Provided the deterioration is confined to the surface of the concrete and there has been no significant corrosion of the reinforcement, the following measures can be adopted; the one actually selected will depend on the circumstances of each case:

(1) The surface of the concrete must be prepared by wire brushing, or water jetting so as to provide a clean strong base on which the new coating will be applied. It is essential to do this preparatory work carefully so that the best possible bond is obtained.

(2) On the prepared surface of the concrete, apply two thick brush coats of grout made with 2 parts ordinary Portland cement and 1 part styrene butadiene latex by weight. The first coat must be well brushed in and the second coat should be at right angles to the first.

OR

(3) Apply to the prepared surface of the concrete, two coats of epoxide or polyurethane resin, to a minimum thickness of 0·3 mm. For drinking water reservoirs it is likely that polyurethane would not be acceptable as the coating must be non-toxic, non-tainting and should not support bacterial growth.

OR

(4) Where the attack has been severe and there has been some penetration of water to the reinforcement resulting in limited corrosion, the application of a 50 mm thick layer of gunite, reinforced with a light mesh, would be justified. Detailed information on gunite is given in a later section (p. 151).

Bituminous compounds have in the past been used for this type of protection, but generally have not proved particularly durable. They do not bond well to damp concrete, and some may impart a phenol taste to the water.

The Complete Lining of Deteriorated Structures

Cases sometimes arise, particularly with water towers and similar elevated structures, where the provision of a complete lining is required in order to ensure a satisfactory standard of water tightness. For a complete lining, there are two basic alternatives: insitu material or preformed sheeting.

Insitu materials. These include mastic asphalt, bituminous coatings, cement grout composed of OPC and SBR latex, epoxide and polyurethane resins, and gunite. As previously mentioned, for structures holding drinking water, the coating must be non-toxic and non-tainting.

For the coatings, the correct preparation of the base concrete is essential. The author considers that it is most advisable for the supplier of the material to be responsible for all aspects of the work *i.e.* preparation of the concrete, and supply and application of the coating, as this avoids divided responsibility.

Prior to the commercial use of organic polymers (derived from the petrochemical industry) such as epoxide resins and polyurethanes, waterproofing of the concrete was provided by two or more coats of bituminous emulsion or mastic asphalt tanking.

(a) *Mastic asphalt*. Mastic asphalt has gone out of favour because it required a protective/supporting wall on the water face, so only a brief description will be given here. For more detailed information, readers should refer to the Mastic Asphalt Council (see Appendix 3).

The two British Standards for mastic asphalt for tanking are BS 1097: Limestone aggregates and BS 1418: Natural rock asphalt aggregate. There is no British Code of Practice which deals specifically with the use of mastic asphalt for the waterproofing of water-retaining structures but the relevant clauses in CP 102: Protection of buildings against water from the ground, can be considered as applicable.

The following points are essential for the successful use of mastic asphalt

(i) Both horizontal and vertical asphalt should be restrained by a protective layer to prevent displacement by any hydrostatic pressure.
(ii) The mastic asphalt membrane must be continuous.
(iii) Horizontal surfaces (floors) should be covered with a minimum thickness of 30 mm of mastic asphalt, laid in three coats. Vertical and sloping surfaces should have a minimum thickness of 20 mm, also in three coats.
(iv) The maximum possible bond should be obtained between the base concrete and the asphalt. This can be achieved by removal of laitance by one of the usual methods, followed by removal of all dirt and grit.
(v) Horizontal protection is usually provided by means of a 50 mm thick cement/sand screed. Protection of vertical asphalt can be provided by concrete or clay bricks or blocks, built 40 mm clear of the face of asphalt. The 40 mm void is then carefully filled with mortar, course by course, as the work proceeds.

(b) *Bitumen and bituminous emulsion*. If bitumen based products are used in drinking water tanks, a special grade is required that is non-toxic and non-tainting. It has been known for many years that phenol tastes can arise in water kept in structures lined with bitumen.

The author's experience is that this type of coating is not very effective in the long term, but has two advantages, low first cost, and ease of application, and is therefore useful for temporary work.

The surface of the concrete should be free of dirt and weak laitance and should be as dry as practical. The material is best applied by brush in a minimum of two coats, the second at right angles to the first to help eliminate pin holes.

(c) *Latex grout: polymer emulsions and Portland cement*. The treatment of the floor and walls of a water-retaining structure with a grout composed of two parts Portland cement and one part styrene butadiene resin latex emulsion (SBR) by weight can be effective in sealing the surface of slightly porous concrete subjected to a low head of water. There are no figures available for maximum safe water pressure, but the author would not feel confident in the use of this material under a pressure exceeding 1·50 m head.

As with bitumen, it is simple to apply and relatively cheap compared with more sophisticated coatings. Provided the limitations mentioned above are kept in mind, it can prove a useful solution to minor loss of water from small tanks and channels. There are a number of proprietary latexes on the market, but taking as an example the SBR latex 29 Y 40 manufactured by Revertex Ltd., and sold under proprietary names by various firms the following is a guide to its use:

(i) Mix: 2 kg ordinary Portland cement to 1 kg latex
(ii) Coverage in two coat work: 5 m^2/4 litres of grout
(iii) The surface of the concrete must be clean and free of weak laitance
(iv) Each coat must be well brushed in, and allowed to dry before the next one is applied. The second coat must be applied at right angles to the first

Acrylic resin emulsions can also be used, but they tend to be more expensive.

In addition to the basic emulsions mentioned above there are a large number of proprietary materials on the market; the author feels that the most satisfactory results are likely to be obtained by employing firms who supply and apply the coatings.

(d) *Epoxide and polyurethane resin coatings*. These materials are basically organic polymers and can be formulated to give the coatings a wide range of characteristics such as chemical resistance, bond strength, modulus of elasticity, viscosity, setting and curing time, ability to cure in very wet conditions including complete submergence, abrasion resistance and colour.

The client (or his professional adviser) should specify as clearly as possible the conditions under which the coating has to operate. If the coating has to possess considerable elasticity, this should be defined as accurately as possible. Such expressions as 'the coating must be capable of spanning hair cracks' should not be used as there is no general acceptance of the width of a hair crack. For a coating having a total thickness of 0·3 mm, a crack opening to 0·6 mm means an extension of 200 %.

This type of coating will bond very strongly to clean, good quality concrete, and the bond is an essential factor in the long term serviceability of the coating. When failure occurs it is normally bond failure at the interface with the base concrete. It is essential that all weak laitance and any surface contamination such as oil, etc. be completely removed. It is better if the preparation of the concrete and the formulation, supply and application of the coating, are all carried out by one firm as this avoids divided responsibility. Figure 4.11 shows the application of a polyurethane lining to a sewage tank.

For the complete lining of a liquid retaining structure, a minimum thickness of 0·5 mm is recommended and this would normally require three coats. The resin can be applied by brush or airless spray. When properly executed it provides a completely jointless lining which is very durable.

As with all applied coatings which require to be bonded to the base concrete, the limiting factor may be the actual quality of the concrete. With a weak porous concrete the effective life of the lining may be disappointing, simply because the concrete is not strong enough to hold the coating in position. In such a case, reinforced gunite may be the best solution.

(e) *Reinforced gunite.* Gunite is a pneumatically applied material, consisting of cement, aggregate and water. Depending on the maximum size of aggregate used and the grading, it can be considered either as a mortar or a concrete.

The advantage of this material for watertight linings compared with concrete, is that it does not require formwork, is self-compacting, and that except for large structures, or in those cases where the gunite is bonded to the existing structure, all the joints are monolithic. In addition to providing a watertight lining, reinforced gunite can be designed to strengthen an existing structure. This use of gunite has been described in Chapter 3.

Gunite is used more extensively in the USA, South Africa and Australia than in this country. However, in recent years its use in the UK for new swimming pool shells and the lining of deteriorated pools and other liquid-retaining structures, has increased.

Gunite work is usually offered as a 'Design and Construct' package deal type of contract, by a few specialist firms. With certain aggregate gradings and the use of specialist 'gunning' equipment, equivalent 28-day cube strengths in excess of 50 N/mm^2 can be obtained compared with about 35 N/mm^2 for normal quality water-retaining concrete. The higher compressive strength is accompanied by higher tensile and bond strength as well as impact resistance.

When reinforced gunite is used to line a structure, the reinforcement can be pinned to the old walls and floor and the gunite then forms a monolithic lining bonded to the existing structure. An alternative, which is usually adopted when the existing structure is found to be in poor structural condition (low strength concrete, numerous cracks etc.), is to separate completely the new gunite lining from the old structure. The separation is generally formed by a membrane of 1000 gauge polyethylene sheeting. The reinforced gunite lining then forms an independent structure, and would be designed as such.

FIG. 4.11. Application of polyurethane waterproof lining to water tank (courtesy: Colebrand Ltd).

If a bonded lining is used, then all movement joints in the existing structure should be carried through the gunite. When a liquid-retaining structure has cracked at locations other than along the line of joints, it means that these cracks may act as movement joints. This is the principal reason why an independent unbonded lining is adopted.

It was mentioned earlier in this section that unless the structure is very large, the gunite-independent lining would have no movement joints, but only monolithic day-work type joints. A natural question, therefore, is why should provision have to be made in a concrete lining for thermal contraction (stress relief or contraction joints) while gunite appears to be satisfactory without any movement joints at all. It is not easy to give a completely clear, unambiguous answer to this, but the factors involved are:

(a) A reinforced gunite lining is usually appreciably thinner than a lining in reinforced concrete. The gunite lining has no formwork, resulting in far less build-up of heat from the hydrating cement, and consequently lower thermal stresses, than concrete.

(b) Gunite linings are usually reinforced with heavy steel fabric and the amount of horizontal (distribution) steel is likely to be greater per unit thickness of wall or floor, than would be provided for concrete.

(c) Good quality, high strength gunite, as placed, has a very low w/c ratio, about 0·33 to 0·35. Concrete used for lining a liquid retaining structure would normally have a w/c ratio of about 0·48 to 0·50.

The above factors add up to the fact that there would be appreciably less thermal contraction in the gunite lining than in an insitu concrete lining, followed by less drying shrinkage, coupled with a heavier weight of reinforcement to control cracking.

However, cracking is not unknown in gunite linings, but is generally caused by inadequate precautions being taken to prevent rapid drying-out of the surface under adverse weather conditions (hot sun, strong winds). This rapid drying out shows in the form of fine surface cracks. It is advisable for provision to be made for proper covering up of the gunite as the work proceeds.

The bulk density of high strength gunite is about 0·9 that of normal dense concrete, *i.e.* $0{\cdot}9 \times 2380 = 2100\ \mathrm{kg/m^3}$. The total weight of a gunite lining is likely to be appreciably lighter than a concrete lining and therefore when the structure is subjected to uplift from ground water it is usual to provide pressure relief valves in the floor. This is particularly important with separate unbonded linings.

The surface finish of gunite is generally rougher than that of good quality concrete. This may be considered a serious disadvantage when gunite is used to line potable water tanks as some water engineers consider the rough surface would encourage the growth of algal and fungoid growths. This can be partly overcome by finishing the gunite with a wood or steel float.

FIG. 4.12. Defective swimming pool prior to new gunite lining (courtesy: Gunite Swimpools Ltd).

Figures 4.12 and 4.13 show an old deteriorated swimming pool before and during the application of a structural gunite lining.

Preformed Sheeting

There are three basic methods of lining water-retaining structures with sheet material: fully bonded lining; partially bonded lining (spot bonding); and unbonded lining.

In the UK for new structures such as reservoirs and water towers a fully bonded lining is normally adopted. In recent years, the technique of using a loose 'bag' of PVC sheeting has been introduced for small swimming pools and it could be used for other structures in certain restricted circumstances.

FIG. 4.13. Defective swimming pool being lined with structural gunite (courtesy: Gunite Swimpools Ltd).

(a) *Fully bonded and partially bonded linings.* The basic principles for successful application of these two types of lining are very similar and therefore are considered together. A partially bonded lining is used where appreciable further movement in the structure is anticipated. Care is taken to ensure that the lining is left unbonded across all planes of movement and in this way, excessive build-up of stress in the lining material is avoided. Generally about 75% of the area of lining is bonded to the substrate.

The great advantage of a bonded lining compared with an unbonded one is that should the lining be damaged in a limited number of places, it is most unlikely that leakage would occur. The reason for this is that the liquid cannot travel behind the lining until the adhesive has become very deteriorated. The adhesives used are water-resistant, but it must be admitted that if water does penetrate through the membrane, in the course of time the adhesive will suffer deterioration. The degradation of the adhesive and consequent loss of bond will show itself by the bulging of the lining and this should be detected in the course of normal maintenance inspections before the water is able to penetrate through the structure.

The disadvantage is the careful preparation which is required of the base concrete; this preparatory work is detailed below. However on balance, the author is of the opinion that for structures such as water towers etc. the advantages of a fully or partially bonded lining far outweigh the disadvantages.

The material mostly used is polyisobutylene (PIB). Some general information on this material has been given in Chapter 1.

As with any material requiring good bond, careful preparation of the base concrete is essential. In many cases, linings of this type are applied to old structures which over the years have been coated with other materials (such as bitumen) to help improve watertightness. All such coatings which are unsound or blistered must be completely removed. It is important that the surface to which the PIB sheeting will be bonded, is strong and relatively smooth. A scabbled or grit blasted surface would not be suitable as the irregularities on the surface created by the pieces of coarse aggregate would cause local stress in the sheeting. Therefore if scabbling or similar is required to remove contamination a thin fully bonded coat of rendering must be applied to the base concrete. If the concrete is reasonably clean, but contains rough areas, these must be ground down.

The sheeting should be well bonded to the concrete with a special adhesive which is not water-sensitive and all seams must be lapped and solvent (cold) welded. An alternative is to use a special tape which is inserted between the sheets and then the joint is fused.

Figure 4.14 shows a previously leaking reservoir after lining with fully bonded PIB sheeting.

It is important that the sheeting be continuous over the floor, and carried up columns and walls to at least 300 mm above top water level. If the sheeting is finished below water level, then it is likely that continuous immersion of the unprotected edges over a long period of time will result in deterioration of the adhesive, causing loss of bond and allowing water to penetrate behind the lining. Even so, the top exposed edge of the sheeting needs careful finishing to avoid ingress of condensation as this is always present in tanks holding liquid at ambient temperature. Reservoirs and water towers contain a number of pipes, both inlet and outlet, and the PIB must be carefully cut and fitted around these and all joints lapped and

FIG. 4.14. View inside reinforced concrete reservoir lined with fully bonded PIB sheeting (courtesy: Gunac Ltd).

solvent welded. Any joints in the structure need special attention and detailing so as to prevent build-up of stress in the lining.

The lining of tanks with PIB is highly specialised and should only be entrusted to an experienced firm who should be required to give a guarantee of satisfactory performance for not less than 10 years.

Preformed sheeting is fairly easily damaged by small metal tools, hob-nail boots etc. and therefore special precautions have to be taken when carrying out inspections, maintenance and cleaning work.

(b) *Unbonded lining*. The use of an unbonded lining of flexible sheet material for the waterproofing of water retaining structures appears to have started with small private swimming pools. The material generally used is polyvinylchloride (PVC) (also known as vinyl), and the sheets vary in thickness from 0·80–1·5 mm. It is unlikely that PVC would be accepted for lining potable water tanks due to possible leaching of chemicals from the PVC.

The advantages claimed for the use of unbonded lining are that movement of the structure does not induce stress in the lining, and that provided the surface on which it is laid is reasonably smooth, no special preparation is required. Also, with open structures (not roofed), the sheeting can be laid in almost any weather. The author has seen this method used on roofs and for lining small swimming pools.

While the horizontal area which can be covered by unbonded sheeting is virtually unlimited, the vertical height must clearly be limited, probably to a maximum of about 1·5 m, although the author has not seen any figure quoted. With any form of complete lining, it is important that it should not be subjected to back-pressure from ground water when the structure is empty as there would be a danger that the sheeting would be forced away from the substrate. In the case of unbonded linings it is generally considered essential that all necessary precautions be taken to prevent any back-pressure developing, however slight. The weight (pressure) of the contained liquid moulds the sheeting to the exact shape of the structure, and if at a later date, the lining is forced out of position by pressure from behind, it is unlikely ever to return to its previous position and shape.

FIG. 4.15. View of badly deteriorated industrial settling tank prior to strengthening and repair (courtesy: Cement Gun Co. Ltd).

FIG. 4.16. View of settling tank in Fig. 4.15 after preparation of concrete and during fixing of new reinforcement prior to guniting (courtesy: Cement Gun Co. Ltd).

FIG. 4.17. View of settling tank in Fig. 4.15 after completion of guniting (courtesy: Cement Gun Co. Ltd).

The External Use of Reinforced Gunite to Strengthen Liquid-Retaining Structures

A previous section has dealt with lining liquid-retaining structures with reinforced gunite. This type of lining is usually applied to below-ground structures. For structures which are above ground, such as many of the tanks at sewage treatment works, sludge and settling tanks for industrial effluent, and water towers, it is sometimes necessary to strengthen the tanks from outside. Reinforced gunite is very suitable for this work.

The gunite coating, reinforced as required to provide the necessary additional strength, is fully bonded to the base concrete. The existing concrete must be carefully prepared to receive the gunite. It is good practice to cut away all spalled and defective concrete, remove rust from reinforcement, cut out and repair all major cracks, and thoroughly clean the surface of the concrete by grit blasting or high velocity water jetting. The new reinforcement is then securely pinned to the old concrete and gunning is commenced.

Figure 4.15 shows a badly deteriorated concrete settling tank used for industrial effluent. Figure 4.16 shows the same tank being prepared for guniting. Figure 4.17 shows the same tank after completion of external structural gunite.

4.4 REPAIR WHERE LEAKAGE IS INWARDS INTO THE STRUCTURE

So far in this chapter consideration has been given to the repair of liquid-retaining structures where the leakage is outwards from the structure. Leakage or seepage into the structure can take place with both liquid-retaining and liquid-excluding structures. In the former type the infiltration would only occur when the pressure outside exceeded the pressure inside, in other words when the structure is partially or completely empty. When infiltration does occur, then this can be difficult to cure completely, because it is usually impossible to carry out the repair from the water (outside) face.

4.4.1 Repairing Leaks in Basements

Consideration of this problem includes basement structures in which special electronic equipment is fixed, and no moisture penetration can be tolerated. In cases of this category, it is usual to tank the structure, *i.e.* provide a complete waterproof membrane under the whole of the floor and carry it up the external walls to above ground level. Even so, the problem of ensuring that such a large area of membrane is completely watertight and remains so during the construction of the structural floor and walls and for the lifetime of the building, is a formidable one, and is not always successful.

The author was asked to advise on the waterproofing of a basement 15 m deep with a water table rising to 10 m above the floor; no moisture penetration could be tolerated. The final solution involved the construction of a circumferential retaining wall with an 'open area' between this wall and the external walls of the building. In addition, maximum precautions were taken to ensure that the membrane below the floor was very durable. The membrane below the floor consisted of flexible PVC sheeting 1·5 mm thick, with solvent welded seams, carried across the 'open area' and up the inside face of the reinforced concrete retaining wall to 1·00 m above the floor of the area and turned into a specially formed groove in the concrete. The PVC membrane was itself protected against damage during the laying of the structural floor slab by a further sheet of PVC, 1·00 mm thick; the seams of this protective sheeting were only spot welded to hold them in position. On top of this protective membrane a 25 mm thick cement sand screed was laid.

The above brief description is a special case where the achievement of complete watertightness to ensure total exclusion of moisture penetration, was of sufficient importance to justify the very high extra cost involved.

It is obvious that at the design stage, detailed consideration should be given to the standard of watertightness required, bearing in mind that good quality, well compacted concrete is watertight but not vapourtight. The standard required for the sump of a sewage pumping station would be different to that of an underground car park, which in turn would differ from a basement used for storage of materials liable to be damaged by high relative humidity, or the type of special structure referred to above.

When the specified standard of watertightness has not been achieved and repairs are required, the problem is to decide how these repairs should be carried out, taking into account the circumstances of each case as far as they are known. The words 'as far as they are known' are important because in most cases nothing is known for certain about the actual condition (impermeability) of the concrete which is behind the inner face of the wall. All that is known is that water penetration is occurring; this may be along joint lines or it may be in random areas of the floor and walls.

In the case of penetration on joint lines, this may be due to a fractional opening of a 'monolithic' joint which does not have a water bar and/or sealing groove on the outside. If there is a water bar then the seepage indicates either some displacement of the water bar or honeycombed (undercompacted) concrete around the bar. In the random areas, this is due to undercompacted concrete.

What is emphasised here is that the extent of the undercompaction (or degree of porosity) is not known, unless cores are taken or an ultrasonic pulse velocity survey is carried out. It is the author's experience that it is very unusual for either of these two methods of investigation to be adopted. It is general practice to stop the leaks by sealing the inside face of the floor or wall. There are a considerable number of proprietary materials on the market in this country, the Continent, USA etc., which when properly used, will seal off the inflow of water. Most of these materials are fairly new inasmuch as they have been developed over the last 15–20 years. Their durability in terms of the normal life time of a reinforced concrete structure is therefore not known for certain. This statement is not intended to suggest in any way that these or other new materials should not be used. If this attitude were adopted, there would be no progress at all in the development of new materials and techniques.

Questions are sometimes asked whether the water penetration and consequent repairs will result in increased maintenance costs. It is the author's opinion that it is reasonable to assume that some additional maintenance expenditure will be required compared with the same structure if no water penetration had occurred at all. However, it would be quite unrealistic to assume that a large reinforced concrete basement can be constructed in a subsoil which contains a water table, without any water penetration at all. It can be done in theory, but not in practice unless the basement is tanked. From this, a fundamental question arises, namely, what effect will water penetration have on the long term durability of the structure from the point of view of structural stability and maintenance. The author has already given his opinion on maintenance costs, but he feels that detailed consideration of the effect of water penetration would be useful.

The reinforced concrete walls of basements are either (a) load-bearing or (b) panels spanning between reinforced concrete columns. The floor slab is usually uniformly supported on the ground with additional reinforcement to take any uplift due to water pressure, or is a suspended slab, also with additional reinforcement. The first matter to be considered is whether the ground water will attack the concrete itself. Some ground waters are mildly aggressive to Portland cement concrete, but not sufficiently so as to cause any significant attack on an adequate thickness of high quality, dense impermeable concrete. When sulphate-resisting Portland cement is used, it is always emphasised that a good quality well-compacted concrete is also required, so that the sulphate-resisting properties of the cement will operate to the best advantage.

It has been stressed several times before in this book that steel embedded in Portland cement concrete is protected from corrosion by the intense alkalinity of the cement paste. In other words the steel is passivated. Unless this passivation is broken down either by a reduction in the alkaline environment provided by the cement paste or by other factors such as the presence of chloride ions, corrosion of the steel will not occur.

If the leak is sealed off on the inside face of the wall or floor, this does not prevent water entering the concrete from the other side, but it does stop any flow-through. This prevention of flow and the establishment of static conditions is more important than may appear at first sight. Concrete subjected to continuous water pressure will in the course of time become saturated. The rate at which the water will penetrate the concrete will depend on the permeability of the concrete. With high quality, dense, well compacted concrete the permeability rate will be very low. This very slow passage of water into the concrete will not as far as is known at present, result in deterioration of either the concrete or the steel reinforcement, unless the water contains aggressive chemicals in solution.

Clearly, the actual amount of water which penetrates into the concrete is an important factor. Unfortunately in the type of structure considered here, this factor is not known. If for example, the concrete in the outer part of the wall or lower part of the floor slab were very porous, there may be so much water penetration as to reduce the alkalinity of the cement paste below the level required for effective passivity and then corrosion of the outer (or lower) bars may occur. However, unless the areas of seepage were considerable, it is unlikely that the strength of the wall or floor as a whole would be affected to any significant degree.

From this brief discussion it is obvious that each case of water penetration must be considered on its merits. Surface sealing of walls and floors on the inside face, remote from the point of entry of the water, has been successful in providing a reasonably dry basement. It can therefore be considered as an established and accepted method of repair. An alternative or additional method is pressure grouting, which has been briefly described earlier in this chapter. It cannot be relied upon to completely seal off areas of infiltration, but if carried out by experienced contractors, it will greatly

FIG. 4.18. Deteriorated and leaking basement (courtesy: SIKA Contracts Ltd).

FIG. 4.19. Basement in Fig. 4.18 after waterproofing to walls and floors (courtesy: SIKA Contracts Ltd).

reduce the penetration of water. It has the advantage that the grout will penetrate into the sections of concrete which are honeycombed, thus providing direct protection to the reinforcement in that part of the wall or floor.

It is the author's opinion that a specification for a new structure of the type considered here, should contain clear directions on how any necessary repairs to prevent ingress of water should be carried out. Such a specification may require both pressure grouting and surface sealing when the wet area or amount of moisture penetration exceeds stated figures as this would help to ensure long term durability.

Figures 4.18 and 4.19 show an old deteriorated and damp basement before and after repair.

4.4.2 REPAIRING LEAKS IN ROOF SLABS

In some structures, minor leaks in the roof may not be viewed with much concern apart from the danger of corrosion of the reinforcement. However in the case of drinking water reservoirs any leak is a potential source of contamination. Joints and cracks in the concrete slab are likely to be the principal cause of leakage. Porous concrete may contribute, but is seldom the main cause of the trouble.

It is now normal good practice to provide a waterproof membrane over the whole of the roof slab, but this was not the case some 30 years or so ago. The absence of the membrane is often accompanied with inadequate falls to the slab. This results in 'ponding' and may lead to the gradual saturation of the concrete. In the course of time the alkaline environment around the steel may be reduced to such a level that corrosion occurs.

Where there are no definite leaks, but rust stains on the soffit of the slab, it is very difficult to decide whether the corrosion is due to the porosity of the concrete cover, *i.e.* the soffit concrete, or to penetration of water from above.

The earth cover, if any, must be removed, and the surface of the concrete thoroughly cleaned, with special reference to joints and cracks. In repairing joints and cracks, the major decision has to be taken on the method of repair, whether to use a rigid or flexible material. With uncovered, exposed slabs, the temperature range may be 40–50 °C, while with a cover of 300–400 mm of earth, the range may not exceed 15 °C. With a large temperature range, the joints and/or cracks are likely to open and close seasonally and therefore flexible sealants should be used. With a much smaller temperature range, many of the joints and cracks can be safely repaired with a rigid material, so as to 'lock' the joint or crack.

Many older structures have no purpose-made movement or partial movement joints in the roof slab. This often results in some of the construction joints opening and forming what in reality is a number of stress relief joints. Cracks are sometimes formed by the same cause. Where there is a definite leak through a joint, a practical way to effect the repair is to remove the whole of the existing sealant and replace it with new. If there is any inert filler in the joint this should also be renewed. The new sealant can be any of the materials described in Chapter 1, bearing in mind the characteristics of the various types. If preformed neoprene is selected, the width of the gasket must be wider than the groove into which it will be fixed, and the use of the correct 'oversize' is essential if a watertight joint is to be achieved. The neoprene channel section sealing strip shown in Figure 4.20 is new in the UK but it appears to have considerable potential. Figure 4.21 shows the use of PIB sheet and an insitu sealant. If a crack is sufficiently straight it should be possible to seal it with either of the methods shown in Figures 4.20 and 4.21. Unfortunately cracks are seldom straight, and then the only practical method is to cut it out with the special tool shown in Figure 4.22, and after cleaning out grit and dust etc. fill it in with an insitu sealant.

In structures built in the 1920s and 1930s, it is sometimes found that the roof slabs are reinforced with XPM instead of round bar reinforcement as is the practice today. The author has found that quite severe corrosion of XPM can occur without spalling or cracking of the concrete. Rust stains are visible and sometimes the outline of the XPM can be clearly seen. When this happens the concrete should be removed to allow the XPM to be examined and an assessment made of its value as tensile reinforcement. If the XPM is seriously corroded, it is advisable to provide new reinforcement on the soffit of the slab, anchored into the beams and properly gunited in. At the same time a decision should be taken as to the cause of the corrosion, *i.e.* penetration of water from above or below. In practice it may not be possible to arrive at a clear cut answer to this, and in this event, it would be prudent to provide protection on the top surface and on the soffit of the slab.

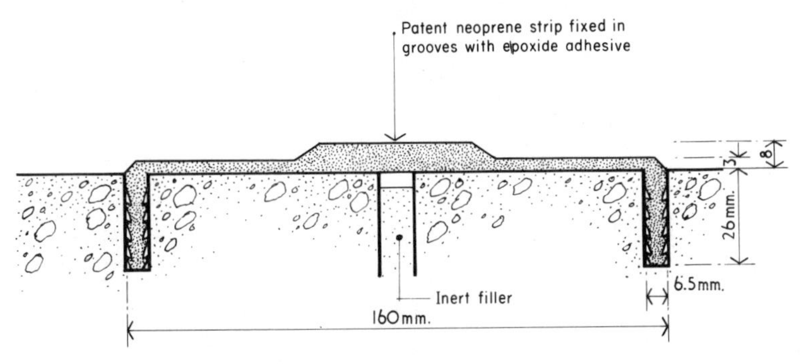

FIG. 4.20. Method of sealing joint in R.C. roof slab using channel-section neoprene strip.

If the suspected seepage is widespread, the provision of a new fully bonded waterproof insitu membrane of polyurethane, or preformed sheeting of polyisobutylene, or 1000 gauge Bitu-Thene would be justified. An alternative is the use of unbonded PVC sheeting. The advantages of this latter method are that the sheets, being unbonded, are not subjected to strain

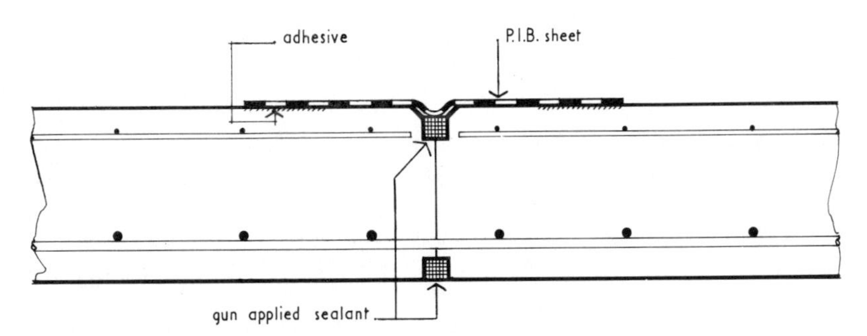

FIG. 4.21. Method of sealing joint in R.C. roof slab using PIB sheet and insitu sealant.

FIG. 4.22. Random crack being opened up by an Errut-MacDonald machine (courtesy: Errut Products Ltd).

due to movement of the roof slab; also, they can be laid in almost any weather.

However, if the unbonded sheets become damaged and water can pass through the holes or tears, then it will gradually flow over the roof slab until it finds a weak spot and then will penetrate the slab. With fully bonded sheets this will not occur as long as the adhesive remains intact.

Roofs of reservoirs are usually very exposed and it is essential that the unbonded membrane be held down against the suction which develops during periods of storm and strong wind. This is usually done by carefully spreading rounded (not angular) shingle on the surface to a depth of about 50 mm.

The sheeting is in fact fixed to the perimeter of the roof and is carried up and fixed against parapet walls, pipes and other members which pass through the roof slab. The sheets are lapped and solvent (cold) welded and then finally sealed with a special material. PVC sheeting is very tough and durable, but it must not be in contact with bitumen, creosote, tar or similar compounds. Detailed information on the use of insitu coatings and preformed bonded sheeting for lining reservoirs and similar structures has been given earlier in this chapter.

Other materials which can be successfully used to hold-down and protect unbonded sheeting, include no-fines concrete 75 mm thick, and 50 mm thick precast concrete slabs. These two materials can also be used to provide a protective cover to insitu coatings and bonded preformed sheeting, capable of taking foot and light vehicular traffic.

There will be cases where a membrane has been provided, but in spite of this, the roof leaks. The location of the defect(s) in the membrane can be very difficult if not impossible because it is extremely unlikely that points of visible leakage on the soffit of the slab will coincide with the defects in the membrane. The decision on the best method of repair will depend largely on the extent of the leakage. If the leakage is extensive probably the most satisfactory repair will include the removal of the existing membrane and its replacement by new material. At the same time the opportunity should be taken to seal the defects in the concrete slab. If the leakage is relatively small it can be repaired on the underside of the slab by one of the methods described in the previous section.

4.5 GENERAL REPAIRS AND WATERPROOFING OF TUNNELS AND PIPELINES

For the purpose of this section tunnels and underground pipelines will be considered as a single type of structure, similar to the Oxford Dictionary definition of tunnel, namely, 'an artificial subterranean passage'. The problems of repairing and waterproofing large diameter pipelines are similar to those met with in tunnels.

The materials and techniques which have to be used in the waterproofing of tunnels and pipelines are in many ways different to those adopted for waterproofing structures such as reservoirs and sewage tanks. Tunnels and pipelines must inevitably be waterproofed on the inside and against the inflow of water, while as discussed earlier in this chapter, repairs etc. to water-retaining structures can often be carried out on the water face.

Tunnels and large diameter pipelines are usually driven or laid through water-bearing ground and the inflow of water, particularly in the case of tunnels, can be very considerable. Most consulting engineers realise this and accept a certain amount of infiltration. Many papers have been written about tunnels and tunnelling, and also about main intercepting sewers, but it is rare to find this basic problem of watertightness discussed, and details given of the amount of infiltration accepted by the designers. An exception to this is the paper by Haswell on the Thames Cable Tunnel, read at a meeting of the Institution of Civil Engineers in London on 17 February,

1970. The author's experience with main sewer pipelines is that infiltration is a bigger problem than outward leakage.

There is no Code of Practice specifically for tunnels, but CP 2005: Sewerage, does mention infiltration, and this may be summarised as follows:

> 'The rate of infiltration is dependent upon so many factors that a guide to its permissible extent cannot be given and will depend on the judgement of the engineer.'

In view of the importance of infiltration on the satisfactory functioning of a sewerage system, it is surprising that the Code Committee could not be a little more specific and offer advice to engineers who have the professional responsibility for the design including specification for the construction of trunk sewers.

In the case of tunnels, the amount of infiltration will depend on many factors including the pressure of the ground water, the type of lining, whether single or double skin, the materials and techniques used for caulking the joints between the segments and how leaks elsewhere are dealt with. Haswell in his paper quotes acceptance figures of 5·5 litres/m^2/day, and 32 litres/m^2/day in different sections of the tunnel.

In addition to the tunnel lining it is usual to pressure grout behind the lining as work proceeds. The joints in cast iron segments are usually caulked in lead while with precast concrete segments, the caulking material is normally an asbestos-cement based compound.

None of these are 100% effective and so additional waterproofing is required on a certain percentage of the joints. In addition, there is always some leakage at intermediate places. To seal these leaks against the inflow of water, ultrarapid setting compounds have to be used. There are many proprietary materials on the market. The older ones were usually based on Portland cement with an admixture of gauging liquid to promote an almost instantaneous set. Portland cement and HAC in about equal proportions will give a flash set. In recent years organic polymers, such as epoxide, polyurethane, polyester, and acrylic and styrene butadiene resins have been introduced for this waterproofing work. Sometimes they are used in addition to the older materials. The formulators of the polymers can produce compounds with special characteristics tailor-made for specific site conditions.

Underground pipelines for main sewers are almost entirely reinforced concrete; asbestos-cement is also used in the UK for diameters up to about 1·20 m. Asbestos-cement pipes up to 2·0 m diameter are made and used on the Continent and the USA. Both types of pipeline have flexible joints formed with rubber rings. When correctly installed these form a watertight joint, but if the rings are displaced during pipelaying or grit gets between the rings and the pipe, leakage will occur. It is usually at joints in the pipeline and the junction between the pipes and manholes that infiltration is most likely to occur.

With large diameter pipes, 900 mm and over, men can enter the line and inspect and repair the joints. When there is a considerable inflow of water, repair can be difficult and time consuming. It is impossible to remove the jointing rings and to correct their position. The joint has therefore to be

FIG. 4.23. Infiltration through joint in large diameter pipeline prior to repair (courtesy: Colebrand Ltd).

made watertight by the insertion of a sealing compound into the narrow space between the end of the spigot of one pipe and the inside of the back of the socket of the next. Sometimes due to excessive ground movement, or incorrect laying, this space may be extremely narrow in one half of the joint and very wide in the other. Such conditions require the concrete to be cut back so that there is an adequate space for the sealant.

In one particular job where about one thousand such joints had to be made watertight, the prescribed dimensions of the sealing groove were 30 mm wide and 40 mm deep. The main items of work that had to be done were:

(a) The joints had to be cut out and prepared to the dimensions given above.
(b) The inflow of water then had to be completely sealed off by means of an ultra-rapid setting compound.
(c) On completion of (b), the joint surfaces had to be prepared to ensure maximum bond with the selected sealant.
(d) The sealant, a specially formulated flexible polyurethane compound, was inserted and trowelled off flush with the inside of the pipe.

Figures 4.23, 4.24 and 4.25 show the infiltration through a joint, and work in progress to seal the joint.

FIG. 4.24. View of joint in Fig. 4.23 after initial sealing and prior to completion of repair (courtesy: Colebrand Ltd).

With pipelines having diameters less than about 900 mm, the tracing of infiltration points and other defects can only be carried out by closed circuit television. However, it is not possible to assess by photographs just how much damage has been done to a pipeline by chemical attack, in terms of the thickness and soundness of the remaining concrete.

FIG. 4.25. Final sealing of joint in Fig. 4.23 with flexible polyurethane compound (courtesy: Colebrand Ltd).

Special equipment is available for coating the inside of small diameter pipelines and sealing joints. Figure 4.26 shows the inside of a sewer where serious infiltration is taking place through a defective joint. The photograph was taken by closed circuit television.

In the larger cities there are hundreds of miles of brick sewers much of which are 100 or more years old. These sewers were extremely well built, but in spite of this, due to the great increase in the weight and volume of traffic passing on the surface, serious structural faults have sometimes appeared. This means that the sewer must either be relaid or provided with an internal

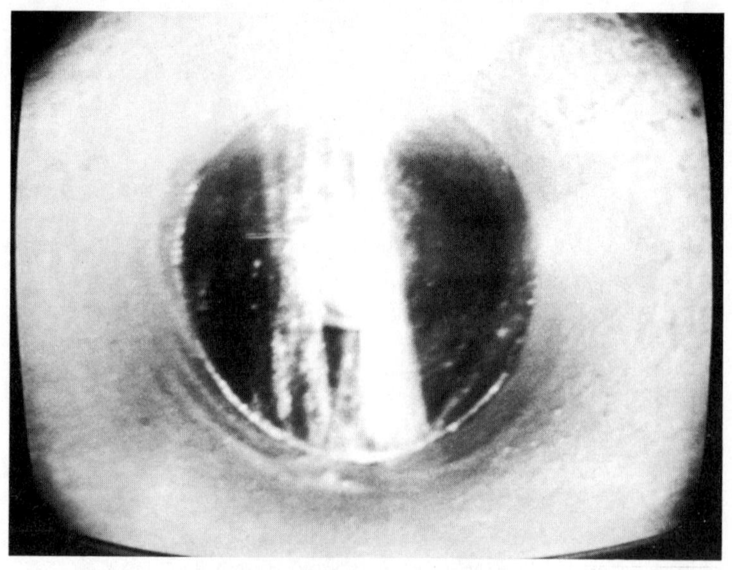

FIG. 4.26. Infiltration through defective joint in 300 mm diameter sewer, view by TV (courtesy: Seer TV Surveys Ltd).

FIG. 4.27. Mock-up of defective brick sewer with new reinforced gunite lining (courtesy: Cement Gun Co. Ltd).

FIG. 4.28. Large diameter old and defective brick sewer before repair (courtesy: Rees Group).

FIG. 4.29. Provision of new structural lining to the sewer shown in Fig. 4.28 (courtesy: Rees Group).

structural lining. A very suitable material for this type of lining is reinforced gunite. Figure 4.27 shows a mock-up of an old egg-shaped brick sewer with a new gunite lining. A test load of double the estimated maximum external load was applied but produced no distress in the strengthened sewer. Figure 4.28 shows a large diameter old brick sewer in urgent need of repair, and Figure 4.29 shows the same sewer being provided with a structural lining.

FIG. 4.30. Large diameter sewer lined with epoxide resin (courtesy: Colebrand Ltd).

However, in many cases the sewer as a whole is structurally sound, but the mortar joints have deteriorated, resulting in increased infiltration. Unless repairs are carried out within a reasonable period, the brickwork will become unsound and the sewer structurally weakened. Defects of this type can be satisfactorily repaired by first sealing off the infiltration through the joints and then repointing them with a styrene butadiene based latex/cement mortar. If the brickwork is itself saturated with ground water, a two-three coat application of a suitably formulated epoxide resin should greatly improve the watertightness of the sewer. A sewer treated in this way is shown in Figure 4.30.

Some brief information on the waterproofing of two tunnels may be of interest.

The Mersey Tunnel. The tunnel lining consisted of precast concrete bolted segments with an inner sheet steel skin. It was hoped that any flow through the structural concrete lining would be sealed off from entering the tunnel proper by the welded sheet steel lining. However, in spite of the most careful welding and supervision, many of the welded joints leaked. These were sealed with a specially formulated epoxide putty and the whole finished with a hot-sprayed epoxide coating. Fig. 4.31 shows the work in progress.

FIG. 4.31. Waterproofing in the Mersey Tunnel (courtesy: Mott Hay and Anderson, Consulting Engineers, and Colebrand Ltd).

Main intercepting sewer. A major part of a large project was in a tunnel and in one section it was found that the infiltration was approximately 67 000 litres (67 m^3)/day. The tunnel lining was precast concrete bolted segments, 3·35 m in diameter. The joints were caulked with a conventional cold caulking compound and the precast lining was pressure grouted behind in the usual way. In spite of this, the infiltration was mainly through the joints between the segments, with a small amount through fine cracks and a few areas of damaged concrete. The work was carried out by a specialist contractor who sealed the joints with a flexible polyurethane of his own manufacture which had been formulated to bond to wet concrete and cure in presence of water. This work reduced the inflow by about 95 %.

4.6 REPAIRS TO CONCRETE DAMAGED BY CAVITATION

Some brief information on the causes of cavitation damage to concrete was given in Chapter 2. Generally cavitation damage occurs on the surface of spillways, aprons, energy dissipating basins, penstocks, syphons and tunnels, which carry high velocity water. There is no agreed figure for the critical velocity below which cavitation damage will not occur, but a figure of about 15 m/s has been quoted. Damage to concrete caused by cavitation can be fairly easily distinguished from normal erosion, as the damaged areas have a jagged appearance while erosion by grit laden water results in a comparatively smooth and/or rounded surface.

From published information on the subject of repairs to cavitation damage, it appears that unless the basic cause of the trouble can be removed, and this is very seldom possible, no repair method yet devised will guarantee lasting success. It is hoped that the following suggestions will prove useful.

1. All defective and damaged concrete must be removed. It is essential that the new concrete or mortar should bond very strongly to a sound high strength base. This may involve cutting away a considerable quantity of concrete. The new concrete or mortar must be high strength; as stated previously in this book, attempts to bond a high strength topping to a weaker base will almost certainly result in failure.
2. The concrete used for repair should have a minimum cement content of 400 kg/m^3, unless the thickness of the new concrete exceeds about 750 mm when the cement content can be reduced to 360 kg/m^3, a maximum w/c ratio of 0·45 and the aggregates should be a high quality crushed rock or flint gravel with a well graded clean concreting sand. Full compaction of the concrete, followed by adequate curing, is essential. Special care must be taken to provide as smooth a surface as possible, and there must be no lipping with adjacent concrete. Any high spots which remain should be ground down, and depressions filled with a fine epoxide mortar.
3. The junction of the old and new work is particularly vulnerable as a fine shrinkage crack usually appears around the perimeter of the new concrete. It is advisable to wire brush a band of a total width of about 300 mm (150 mm on the new concrete and 150 mm on the old), to remove all weak laitance, and then to apply two coats of epoxide resin.
4. Usually the repairs have to be carried out as quickly as possible, but it is important that speed should not be allowed to adversely affect the quality of the work. Consideration can be given to the use of ultrarapid hardening Portland cement or HAC. In both cases the

minimum cement content should be the same as for ordinary Portland (as recommended above), but with HAC, the maximum w/c ratio should be 0·40. Advice on the use of both these cements should be obtained from the manufacturers; the reader should also refer to Chapter 1.

5. If the damaged area is extensive, but of shallow depth, consideration can be given to the use of high strength gunite having a minimum thickness of 75 mm. Depending on the area, it is usual for the gunite to be reinforced with a light galvanised fabric. The gunite should be specified to have a minimum equivalent cube strength at 28 days of 55 N/mm^2. Gunite work is highly specialised and should only be entrusted to an experienced firm.
6. Experience in the USA on repairs of spillways and stilling basins has shown that the provision of a sprinkle metallic finish to high quality concrete gives improved resistance to abrasion and cavitation. This type of finish is referred to in more detail in the next section.
7. Recent work in the USA suggests that polymerised fibre reinforced concrete may be particularly resistant to both cavitation and abrasion. Severe damage was caused to the spillway of the Dworshak dam in Idaho and after extensive laboratory research, the Corps of Engineers will use fibre reinforced concrete which is polymerised insitu for the repair of the most badly damaged areas. Brief information on polymerised concrete is given in Chapter 1.

4.7 REPAIRS TO CONCRETE DAMAGED BY WATER CONTAINING GRIT

The factors which are important in determining whether fast flowing water will or will not abrade the surface of concrete were summarised in Chapter 2. Of the four listed factors, probably the most important is the quality of the concrete. The appearance of concrete damaged by grit laden water is quite different to that caused by cavitation. In channels and pipes the wear may be confined to bends, or it may extend along the invert for comparatively long lengths. Really good quality concrete is surprisingly resistant to abrasion. Surface water sewers laid at very steep gradients down the slopes of Mount Carmel in Haifa showed no sign of damage after more than 15 years. There was only flow in these pipelines during the winter (November to March) but during that period of five months the average annual rainfall was about 600 mm.

On the other hand, in the UK, large insitu reinforced concrete channels used to convey sugar beet mixed with earth, stone and grit, were worn away to a depth of 50–75 mm in about two years. The abrasive conditions in the sugar beet channels were much more severe than in the surface water sewers, but there was no doubt that the quality of the concrete in the precast pipes was much higher than in the insitu channels.

If it is established that the quality of the base concrete is satisfactory and erosion has only occurred in a number of isolated places, then these can be repaired by careful patching with high strength concrete. It is advisable for these patches to be cut out with a saw or high velocity water jet, so as to provide a thickness of new concrete not less than 75 mm. If this minimum thickness cannot be provided, then a high quality cement/sand mortar containing an SBR latex should be used. Even so the thickness should not be less than 30 mm. For thinner patches, an epoxide resin mortar should be used.

Regarding the execution of the work, the general recommendations for the repair of cavitation damage will apply, with the exception that the provision of a very smooth surface is not usually required. However, this will depend on the hydraulic requirements of the structure.

Mention was made under the repair of cavitation damage to the use of a metallic sprinkle finish. This type of finish to industrial floors where high abrasion resistance is required, has been described in detail in the author's book 'Floors—Construction and Finishes'. It is best applied to an adequate thickness of high strength insitu concrete while the concrete is still in the plastic state. The sprinkle finish consists of Portland cement, a metallic aggregate and an admixture, and is spread over the base concrete a few hours after compaction and finishing. The time factor is very important and

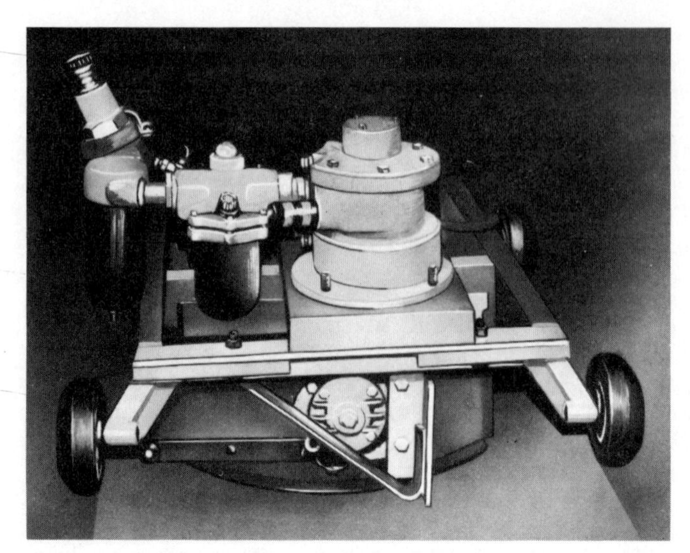

FIG. 4.32. Equipment for laying abrasion resistant invert in concrete pipes (courtesy: Colebrand Ltd).

can only be determined by experience. This thin topping is finished by careful hand trowelling, and then the newly laid area is cured in the usual way. When ferrous metal is used, brown stains will appear; this is caused by the rusting of the fine particles of iron in the surface layer. The rusting has no adverse effect on the durability of the material. The thickness of the metallic finish depends largely on the weight of metal used. With about $5\,kg/m^2$ of metal, the thickness would be about 4 mm. For very severe conditions on industrial floors, $45\,kg/m^2$ of metal with a thickness of 12 mm is sometimes used. These metallic sprinkle finishes are applied by specialist firms, and can be laid on slopes up to about 45° to the horizontal.

The type of repair so far described can only be carried out in locations where there is access for the workmen. This means in the case of pipes, a minimum diameter of 1000 mm. For pipes smaller than this, special equipment and techniques have to be used. Equipment has been developed and used for providing a special abrasion resistant invert to asbestos-cement and concrete pipes. The material used consists of calcined bauxite mixed with a specially formulated epoxide resin. The bauxite is very expensive, and it is likely that carefully selected, graded and washed flint would be satisfactory in many cases. Figure 4.32 shows such an invert being laid in concrete pipes.

BIBLIOGRAPHY

KENN, M. J., Protection of concrete from cavitation damage. *Proc. I.C.E.*, (May 1971), Technical Note 48, pp 5. Discussion, *Proc. I.C.E.*, (Dec. 1971)

CLARK, R. R., Bonneville dam stilling basin. *J. Amer. Concr. Inst.*, **52** (April 1956), pp. 821–837

ANON., Polymerized fibrous concrete to fix dam spillway as research continues on other applications. *Engineering News Record*, Jan. 9, 1975, p. 10

KENN, M. J., Factors influencing the erosion of concrete by cavitation. *Construction Industries Research and Information Association* (*CIRIA*) *Technical Note*, (1968) (1), p. 15

PERKINS, P. H., *Floors—construction and finishes.* Cement and Concrete Association, London, Dec. 1973, p. 132

PERKINS, P. H., *Swimming Pools.* Applied Science Publishers, Barking, Essex, May 1971, p. 358

SCHRADER, E. K. AND MUNCH, A. V., Fibrous concrete repair of cavitation damage. *Proc. A.S.C.E., Journal of Construction Division*, **102** (CO2) (June 1976), pp. 385–402

CHAPTER 5

Repairs to Concrete Marine Structures

It is generally acknowledged that a marine structure is located in a hostile environment and the conditions of exposure are classified as severe. Experience shows that such structures are more liable to damage and deterioration than the majority of land based structures.

Deterioration and damage can occur to a concrete marine structure in many ways which depend largely on its geographical location and site conditions, as well as its design, method of construction and the quality of the concrete. Geographically, the warmer the sea the more rapidly chemical reactions occur; also in areas such as the Red Sea and Persian Gulf, the sea water is appreciably more saline than in the more temperate zones. In the far North and far South the sea freezes for part of the year and the sub-zero temperatures can cause physical disintegration of the concrete. When the moisture in the surface layers turns into ice, the consequent expansion disrupts the concrete. Generally, the most vulnerable part of a structure is the section within the splash zone, that is from low tide to a certain variable height above high tide. The overall height of the splash zone depends on the degree of exposure and the weather conditions.

Site conditions also have a profound effect on the durability of concrete in this class of structure. Some sites are comparatively sheltered while others are exposed to heavy seas and gale force winds. Structures such as sea defence walls and groynes are sometimes subjected to severe abrasion by sand and shingle being dashed against the exposed faces. On the other hand, those parts of a structure which are permanently below low water level are in relatively uniform conditions of temperature, are not subject to alternate wetting and drying and are not exposed to the full fury of waves and wind.

Research into the durability of concrete in sea water has been carried out in a number of countries over many years, and reports on some of this work are included in the Bibliography at the end of this chapter. Professor O. E. Gjorv of Norway has reported on concrete specimens immersed for periods of up to 30 years (Journal of the American Concrete Institute, January,

1971). These tests showed that Portland cements with relatively low C_3A (tricalcium aluminate) content (less than about 8 %), were more durable than similar cements with higher C_3A contents. The tests also proved that HAC was as durable as the best Norwegian and German Portland cements. Even so, a complete understanding is still lacking of the reactions between Portland cement and the salts in solution in sea water, as well as the mechanism of the corrosion of steel reinforcement.

In practical terms, the evidence from actual structures is conflicting. Some structures deteriorate more severely than can be satisfactorily explained, while others show unexpected and remarkable durability. Some of the steel reinforcement in the concrete Mulberry Harbours constructed in 1943 for the Normandy landings, has been found in excellent condition after 30 years, with a cover of only 25 mm. It is clear that the quality of the concrete, in terms of cement content and impermeability, is the dominant factor, and not the thickness of the cover. This fact is confirmed by the durability (freedom from corrosion of the reinforcement) of concrete boats and ferro-cement boats, where the cover is often appreciably less than 25 mm.

The chemical constituents of sea water have been discussed in the section in Chapter 2 dealing with the use of sea water for mixing concrete and mortar.

As far as chemical attack on the concrete is concerned, it is the sulphates in solution which are likely to cause the trouble. In Atlantic water, the sulphate concentration is about 2000 ppm (mg/l) and is largely magnesium sulphate, which is considered more aggressive than the sulphates of calcium and sodium, in equal concentrations. In ground water, a concentration of 2000 ppm would require the use of sulphate-resisting Portland cement in order to prevent sulphate attack on the concrete. However it is well known that Portland cement concrete does not suffer from sulphate attack in normal sea water around the coasts of the UK, provided the concrete is dense and impermeable. This apparent conflict has not been satisfactorily explained, but it is thought that the other salts present have an inhibiting effect on the reaction. It should be noted that the above remarks apply to 'normal' sea water. In some cases the water in estuaries and harbours is contaminated by sewage and trade wastes; also, if tidal currents are obstructed, there may be a build-up of salts in solution as well as aggressive organic compounds. In such cases chemical attack on the concrete can occur.

5.1 GENERAL PRINCIPLES OF REPAIR

A fundamental principle of the repair of any concrete marine structure is that the concrete and mortar used must be of the highest quality in terms of cement content, w/c ratio, compaction and impermeability. There will be in addition, special requirements which will depend on the type of repair, its location, and site conditions.

The basic concrete mix should comply with the following:

Cement. This is generally, ordinary or rapid hardening Portland cement. Sulphate-resisting Portland cement need only be used in special circumstances where the sulphate concentration and/or sea temperature is appreciably higher than in normal Atlantic water. HAC can be used with advantage in cases where a very high rate of gain of strength is required, such as concreting or guniting between the tides. The setting time is similar to that of Portland cements. Further information on HAC and concrete is given in Chapter 1. Although long term tests on HAC concrete immersed in sea water have shown it to be as durable as Portland cement, it is advisable for the manufacturers to be consulted before a final decision is taken to use it.

Cement content. With Portland and HAC the cement content should not be less than 400 kg/m^3 of compacted concrete. In the case of mortar, the mix should not be leaner than 1 part of cement to 3 parts of well graded sand by weight.

Water/cement ratio. With Portland cements, this should not exceed 0·45, and with HAC, it should not exceed 0·40.

Aggregates (from natural sources, complying with BS 882). In some parts of the world, the aggregates are contaminated with salt, often chlorides, and there may be insufficient fresh water available for adequate washing to remove the salt. As previously stated in this book, if the percentage of chlorides expressed as anhydrous calcium chloride exceeds 1·5 % by weight of the cement, corrosion of the reinforcement is likely to occur. Precautions can be taken as described in Chapter 2, in the section dealing with the use of sea water for mixing concrete. However, if the concentration is at, near, or a little above the limit of 1·5 %, the cement content can be increased and this will have the effect of reducing the percentage.

Workability. The workability must be adequate to secure full compaction under the conditions of placing which will exist on site. This may require the use of a plasticiser with Portland cement concrete. There is seldom any need to use plasticisers with HAC due to the fact that it is more coarsely ground than Portland (specific surface of about 2500 cm^2/gm compared with about 3300 cm^2/gm for an ordinary Portland).

Admixtures. Admixtures, apart from workability aids, should generally not be used. With HAC no admixture should be used without the agreement of the manufacturers. In very cold climates, the use of air-entrained Portland cement concrete and mortar can be an advantage in combating the disintegrating effect of freezing temperatures on saturated concrete. Admixtures containing chlorides must not be used with HAC.

Cover to reinforcement. The minimum cover for concrete work is usually required to be 50 mm, but as previously stated it is the quality of the concrete which is more important than the thickness. For high quality gunite, a thickness of as little as 20 mm has been found to be sufficient, but 40 mm should be obtained wherever possible.

When using cement/sand mortar for repairs, the author recommends that the mixture should be applied through a cement gun whenever this is practical, as this helps ensure thorough compaction with a low w/c ratio.

Essentially, the methods and techniques of repair are similar to those described in Chapters 3 and 4, but account must be taken of the severe conditions of exposure and the problems which always arise when working under water or between the tides. Any defects in workmanship or materials will quickly become apparent and deterioration will set in rapidly.

For marine repair work, the author is in favour of using galvanised reinforcement as the galvanising helps to protect the steel and the additional cost is justified. Some information on galvanised reinforcement is given in Chapters 1 and 2. There are many problems associated with repair of marine structures including accessibility, work below water level, work between the tides, and the effect of sudden storms on newly repaired members.

The location, design, and operating conditions of marine structures vary widely and these factors all affect the method of carrying out the repair. For the purpose of this Chapter, the work will be divided into the following sections: Repairs below low water level; repairs between the tides and repairs above high tide level, the 'splash' zone; abrasion of concrete caused by sand and shingle.

5.2 REPAIRS BELOW LOW WATER LEVEL

In most cases repairs below low water level require relatively small quantities of material as the damage usually consists of cracked and spalled concrete to piles, piers etc. However it sometimes happens that fairly large volumes of concrete have to be placed under water to protect foundations against scour; in connection with such repairs, it may be necessary to underpin the structure. There are three basic methods of placing concrete under water: by tremie pipe or bottom opening skip; by bagged concrete; by grout injection of preplaced aggregate.

The following are brief notes setting out the most important factors involved in this type of work. The first, which applies to all three methods, is that work below water level is completely different to work on dry land. It is not a job for amateurs, and should only be entrusted to contractors with the necessary experience and equipment.

5.2.1 Placing by Tremie Pipe

Because of the hazards involved, the strength used in the design should be low, *i.e.* not higher than about 20 N/mm². The slump must be very high, 150–200 mm, which is virtually a collapse slump; at the same time the mix should be sufficiently cohesive so that it will not segregate during transporting and placing. The cement content must be high, not less than 400 kg/m³, and the w/c ratio should not exceed about 0·5. The use of a workability aid (plasticiser) is often necessary. The tremie pipe is usually 150 mm in diameter for 20 mm aggregate and 200 mm in diameter for 40 mm aggregate.

It is essential to organise the work in great detail so that a continuous supply of correctly designed concrete is assured. Once the concrete has started to flow, the end of the tremie pipe must remain submerged to a depth of about 400 mm in the concrete. An adequate number of tremie pipes should be provided; usually one tremie will cope with up to 30 m². When the concrete has to be placed in more than one lift, the surface of the previous lift must be properly prepared by divers to remove all laitance etc. As far as possible, the top surface of each lift should be horizontal.

5.2.2 Placing by Bottom-Opening Skip

It is important that the skip should be filled to capacity with concrete and then lowered slowly through the water onto the concrete previously placed. The weight of the skip and its contents is sufficient to ensure that it sinks below the concrete surface. The skip is then gently raised and the concrete should flow out and into the surrounding plastic concrete. The mix proportions and very high slump are similar to that recommended for tremie work. The placing of the concrete is slower with a skip than a tremie.

Some engineers experienced in the use of both tremie pipes and bottom opening skips consider that the latter are usually more suitable for placing small volumes of concrete.

5.2.3 Bagged Concrete

Bagged concrete, according to the Report of the Concrete Society, No. 52.018: Underwater Concreting, is now only used for very minor and temporary works. However, the author's experience is that it can be very useful for protecting structures from physical damage and from the effects of scour; it can be used in both shallow and comparatively deep water.

The concrete should have normal mix proportions for marine work, namely 360–400 kg cement/m³ of concrete. It is mixed fairly dry, with a w/c ratio of 0·35 or less. The sacks are made of jute, but fine mesh polyethylene is now available and this is very strong and durable. The sacks of concrete are usually laid to bond, similar to block walling.

5.2.4 Grout Injection of Pre-placed Aggregate

Graded aggregate, similar to that used for conventional concrete, is placed in forms or in a prepared excavation. Cement grout is injected at the bottom of the aggregate and rises upward displacing the water. Proprietary equipment and techniques are required for this method, but when it is carried out properly, concrete of excellent quality results. The method can be used with advantage in situations where concrete placing will be difficult, for example in underpinning. But it has been used successfully where placing by tremie and bottom-opening skips was also suitable.

It is normal practice for the specialist contractors who carry out this work to use admixtures in the cement grout, such as pulverised fuel ash, in order to improve the flow characteristics of the grout. The mix proportions of the grout are usually about 1 part ordinary Portland cement to 1·5 to 2·0 parts of clean concreting sand, zones 1 or 2 (BS 882). The w/c ratio should not exceed 0·5. Quality control of the strength of the resulting concrete is difficult and special methods have to be used to obtain realistic test cubes.

Further details of this and other methods of underwater concreting are given in the Report of the Concrete Society previously referred to.

5.2.5 General Principles of Repair Below Low Water Level

As previously stated, the majority of repairs to marine structures involve patching or encasing reinforced concrete members such as piles, piers, etc. To do this below water level using cementitious materials (concrete or mortar) is very difficult. Each job presents special problems and the most satisfactory technique has to be worked out for each project. Cracks which have not adversely affected the strength of the member, and other minor damage can be successfully repaired with epoxide resins and resin mortar. These resins are now formulated so that they can be applied under water. Figure 5.1 shows special equipment for spraying epoxide resin under water.

However, serious structural damage requires repair with concrete or gunite. The first step is to ascertain the extent of the damage. When this is below water level, closed circuit television is very useful in providing information for an initial appraisal of the problem. However, careful examination by divers is usually essential for an accurate assessment of the damage so that a realistic method of repair can be worked out.

If the structure has been in existence for any length of time, the concrete will be covered with marine growths (seaweed, barnacles etc). It will then be necessary to remove these growths completely before the full extent of the damage can be appreciated. Marine growth of all kinds can be quickly and completely removed by high velocity water jets; a short description of this technique is given later in this chapter. High velocity water, using nozzle pressures of about 400 atm (6000 lbf/in^2) has been used for many years for cleaning ships' hulls and its use has now spread to the construction industry for cutting and cleaning concrete.

A typical method of repair to badly damaged piles below water level is to provide a reinforced gunite sleeve. The sleeve is formed in sections and

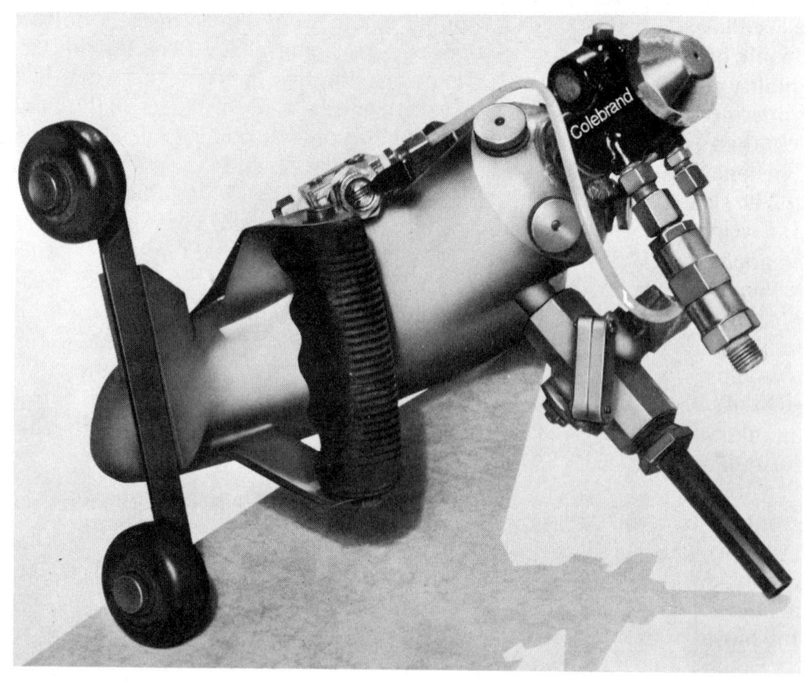

FIG. 5.1. Equipment for underwater application of resin coatings to concrete and steel (courtesy: Colebrand Ltd).

lowered progressively until it reaches its final position; the annular space between the sleeve and the pile is then grouted in.

The author is indebted to the Cement Gun Company for information on an actual job carried out which is summarised below. The damaged members were precast prestressed tubular piles which supported a reinforced concrete platform some 28 m above the sea bed. After a careful investigation (which comprised both closed circuit television and examination by divers) of the damaged piles, it was decided that these would be sleeved from above low water level down to about 1·5 m into the sea bed, which would have to be jetted out for the purpose. Above low water level the piles would be encased direct in reinforced gunite.

A loose sleeve of sheet steel was prepared and fixed in position around the pile above water level, the sleeve being separated from the pile itself by spacers. The shutter (sleeve) was coated with a release agent and wrapped in hessian to ensure a quick and clean release from the gunite shell. A reinforcing cage was then placed in position around the hessian wrapped shutter and high quality gunite applied to a total thickness of about 75 mm.

This allowed about 20 mm of cover to the reinforcement on the inside and 35 mm on the outside. Experience has shown that with properly placed high quality gunite, cover of this thickness is adequate. The gunite was dense and impermeable and could be relied upon to have a compressive strength of not less than 45 N/mm^2. HAC was used for all the guniting work on this job. This enabled a complete unit to be gunned every 24 h.

The sleeves were gunited in sections, the longest being about 9·0 m long and weighing about 6 tonnes. As soon as the gunite had reached a strength of about 30 N/mm^2 the sleeve was carefully released from the shutter and lowered below water level, but with a short section including reinforcement projecting to connect with the next section, which was then constructed in the same way. In this way monolithic sleeves were formed of sufficient length to enclose the whole pile from below sea-bed level to just above low water level. As soon as a full length sleeve had been completed in this way, the annular space between the sleeve and the pile was grouted in with a specially formulated grout which had to displace the sea water and set under water.

Above low-water level the piles were repaired by the direct application of gunite after a reinforcing cage had been fixed around the pile.

The execution of the work was complicated by the tide range of 6·0 m and the strong rather unpredictable tidal currents. The platform had to be kept in operation throughout the repair as it was connected to the main jetty by a 150 m walkway.

FIG. 5.2. Jetty pile damaged above low water level (courtesy: Cement Gun Co. Ltd).

Figure 5.2 shows a jetty damaged above low water level, and Figure 5.3 shows a reinforced gunite sleeve being lowered into position to below low water level.

FIG. 5.3. Reinforced gunite sleeve for repair of jetty pile below low water level (courtesy: Cement Gun Co. Ltd).

5.2.6 REPAIRS WITH EPOXIDE RESINS

Cases occur where some minor damage is caused to piles or other parts of a structure below low water level. In its initial state, the damage may not impair the strength or stability of the structure, but unless it is repaired, deterioration will be progressive and eventually a major and very costly repair, or replacement of the defective member will be necessary.

The damage is usually in the form of cracks (sometimes of unknown origin), spalled arrisses and small damaged areas of concrete. All these greatly increase the chance of sea water penetrating to and corroding the reinforcement, which is likely to result in large scale disintegration of the concrete member. In no circumstances is the old saying 'a stitch in time saves nine' more appropriate than the immediate repair of minor damage to a marine structure.

Regardless of the length of time a member has been immersed in the sea, general cleaning and preparation of the surface of the concrete is normally required. The amount of preparatory work will depend on site conditions and degree of damage. In warm, tropical waters marine growths appear and flourish much more quickly than in the cooler seas.

The type of repair dealt with in this section can now be carried out by specialist firms using modified epoxide resins. These resins can be formulated so as to have many of the most desirable characteristics to suit site conditions. The resins will set and cure underwater and can be applied by airless spray, brush or roller.

For the preparation of the concrete prior to the application of the resin, high velocity water jets are likely to give the best results. If site conditions and/or the size of the repair do not justify the use of the necessary equipment, then general cleaning with wire brushes has to suffice. As with all concrete repairs, the importance of careful and thorough preparation of the damaged surface is essential. It is important that the application of the resin should follow the cleaning operation as rapidly as possible.

In some cases, such as estuaries, docks and harbours, the water may be so turbid that visibility is very poor indeed. This naturally makes repair very difficult, particularly the preparatory work. The resin, in the form of a putty, can be applied by hand ('gloved on') and repairs have been carried out successfully in this way.

Epoxide mortars, composed of the resin and clean, dry, carefully graded silica sand, can be used for repair of damaged arrisses and other defective areas. Generally the mortar should be applied in thin layers, each not exceeding about 6 mm thick. However, if thixotropic fillers are used in conjunction with the fine aggregate, this thickness can be appreciably increased; the actual thickness of each layer should be determined by trial application. In this way the mortar repair can be built up to almost any required thickness. Each layer has to 'set' before the next one is applied. The term 'set' means that the first layer is not disturbed by the application of the next layer. When the work is carried out below water level, this delay can be time consuming and expensive in those cases where speed is important.

Where thicker layers are required, the gunite sleeve method can be used where this is suitable (*e.g.* for such members as piles). For more general application, a specially designed, clamp-on shutter is appropriate. The author is indebted to Colebrand Ltd for the description of a shutter unit which they developed for their underwater repair work. The unit consists of a strong, watertight five-sided box, the edges on the open side being covered with a thick layer of flexible polyurethane sponge. The concrete is prepared in the usual way, but the part of the surface against which the box shutter will be clamped should be brought to a reasonable degree of smoothness. A thick coat of epoxide resin is then applied to the surface of the concrete by brush roller or airless spray. The box is securely clamped into position over the prepared area and the water inside is blown out by compressed air.

Concrete is then pumped into the water-free interior of the box shutter. Provision is made to permit the release of air during the filling operation.

The concrete mix has to be designed with a high slump so as to be virtually self-compacting, and for durability such a concrete will need to contain not less than 450 kg of cement/m^3; the w/c ratio should not exceed 0·50, and 0·45 is better. Plasticisers can be used, but an excessive dosage must be avoided, as this can result in permanent loss of strength. The recently developed 'super plasticisers' could be very useful for this type of work as they would allow the use of a concrete with a low w/c ratio. In this way, repairs to concrete, with concrete, can be carried out below water level. This has many obvious advantages for work below low tide level and between the tides. Some information on super plasticised concrete is given in Chapter 1.

5.3 REPAIRS BETWEEN THE TIDES AND ABOVE HIGH WATER LEVEL, INCLUDING THE SPLASH ZONE

The section between the average levels of low tide and high tide is in reality the lower part of the 'splash zone'. This zone is generally considered to be the one which is subjected to the most severe conditions of exposure. It is in this zone that the greatest fluctuations in moisture content and temperature of the concrete occur. The period during which the concrete is submerged varies. The tide cycle around the coast of the UK is about 12 h from high tide to high tide. Clearly, concrete just below high tide level is totally submerged for the shortest period while that just above low tide is submerged for the longest period. When the concrete is above water level it may be subjected to wind driven spray as well as sand and shingle thrown against it by the waves with very considerable force. The problem of physical abrasion is discussed in the next section.

In regions such as the Persian Gulf and the Red Sea, the sun temperature can reach 60°–65°C and when concrete at or near this temperature is covered with highly saline spray, the spray is likely to be readily absorbed into the surface layers. At night the temperature drops rapidly, and so these changes in temperature plus absorption of salt create conditions likely to result in disintegration of the surface layers of concrete. A similar wetting and drying, heating and cooling cycle occurs in the temperate zones, but within a much smaller range.

At the other extreme of climate, in the far North and South, the sea freezes, also the spray, together with the moisture trapped in the surface layers of the concrete. The expansion of the absorbed moisture can create stresses sufficient to crack and spall the concrete. To mitigate the effect of freezing temperatures, air-entrained concrete is often used. The entrainment of about $4\frac{1}{2}\%$ of air alters the pore structure of the concrete and this is very effective in preventing scaling and spalling under freezing conditions.

The author has not seen any recorded use of air-entrained concrete for marine work other than in very cold climates, but there would appear to be some advantage in its use, particularly where the concrete may be vulnerable to disintegration by the build-up of salt in the surface layers. However, the basic requirements for high quality concrete which have been set out earlier in this chapter should be followed.

Repairs to concrete marine structures in the tidal zone are always a race against time. In no other section of the work is careful and detailed preparation and organisation more necessary. Everything required for the work, materials, equipment and labour must be so prepared in advance that as soon as the tide turns, the next section of the work can be put in hand without delay. The critical section is near low water where the turn-round time is shortest. The usual tide period in the UK is six hours from high to low and six hours low to high. This means that the concrete must be protected to prevent damage to the surface by the incoming tide, which always involves a certain amount of wave action and often fairly strong currents as well. A storm can be disastrous.

For the protection against tidal action (but not storms) of horizontal and sloping surfaces, the use of jute sacks stiffened by an application of cement grout made with about 50% OPC and 50% HAC, can be very effective. This mixture generally produces a flash set. The exact proportions must be found by trial and error as the speed of set will depend on the chemical composition of the two cements actually used. The sacks must be weighted down with large stones or similar to prevent them being moved by the incoming tide. An alternative to the use of jute sacks is to spray the cement mixture direct onto the plastic concrete, followed immediately with a light water spray. This should set almost instantaneously, forming a crust. Further protection in the form of jute sacks, fine mesh polyethylene, or boarding, can then be laid on this protective 'crust'.

Apart from the requirement to provide this initial protection the repair work proceeds in the usual way, but with the utmost dispatch, using the basic principles of mix proportions, cover to reinforcement, and compaction, all as discussed earlier in this chapter.

Where formwork has to be used, as for columns, beams and suspended slabs, then this in itself provides protection until the concrete is mature enough to resist physical damage. The period during which the formwork should remain in position may therefore be appreciably longer than would be required for a land-based structure. It should be remembered that a storm can blow up very quickly, and concrete which has not reached an adequate strength can be badly damaged by the pounding of waves, sand

and shingle. In these circumstances it is not practical, nor desirable to lay down in advance how long the formwork should be kept in position. The specification should contain a flexible clause clearly allowing the Resident Engineer to decide when the formwork can be struck. To be fair to the contractor and thus allow him to price the job realistically, the Bill should contain a provisional item for keeping the formwork in position beyond a certain time.

Consideration can also be given to the use of ultrarapid hardening Portland cement, 'Swiftcrete', and some information on this cement has been given in Chapter 1. The cost is about 50–60 % higher than ordinary Portland cement.

There is a great difference in the rate of hardening of Portland cements and HAC. If HAC is used then it is unlikely that the formwork need be kept in position longer than 24 h after completion of casting. With ordinary or rapid hardening Portland cement it may not be wise to remove the formwork for at least 7 days or even longer. However, the cost of HAC is about four times that of ordinary Portland cement.

It is the author's experience that provided HAC is correctly used, and recommendations on this have been given in Chapter 1 and earlier in this chapter, it is safe to place new HAC concrete against hardened Portland cement concrete in repair work. In the same way, gunite made with HAC can be gunned against existing Portland cement concrete. Gunite made with a Portland cement would require protection in the same way as concrete. This requirement to protect the newly repaired areas for periods up to 14 days in very cold weather, can result in considerable increase in the overall cost of the work.

Mention has been made previously that normal Atlantic water will not be aggressive chemically to good quality concrete made with Portland cement, and the same applies to HAC. There is of course no objection to the use of sulphate-resisting Portland cement; some engineers specify this as an insurance against possible sulphate attack as they obviously feel that the small extra cost is well worth while. For those engineers who agree with this attitude, it should be remembered that the same cement content and low w/c ratio is required, and that some sulphate-resisting Portland cements possess low heat of hydration characteristics. This latter, when it is present may result in rather low strength at early ages of 1–3 days. If the rate of gain of strength is important, then this should be checked with the cement maker before it is finally decided to use a particular sulphate-resisting Portland cement for marine work. Good quality properly made HAC concrete and mortar is rather more resistant to sulphate attack than concrete and mortar made with ordinary and rapid hardening Portland cement.

In estuaries, harbours, and other sheltered positions, the chemical characteristics of the sea water may be significantly different to those in the open sea. It is therefore advisable when considering repairs to marine structures in such areas, to check whether or not the deterioration of the existing concrete is due or partly due, to chemical attack caused by unsuspected concentrations of aggressive chemicals in the water or in the silt and mud through which the piles etc. pass. The lower reaches of many rivers are grossly contaminated with industrial waste. Samples of the water and of the mud etc., which will be in contact with the concrete should be taken at intervals over as long a period as possible so as to provide maximum information on the chemical composition and its probable variation.

5.4 REPAIRS TO MARINE STRUCTURES SUBJECTED TO ABRASION BY SAND AND SHINGLE

The types of structures most frequently affected by abrasion are coast defence works such as sea walls, promenades, aprons and groynes. Figure 5.4 shows serious abrasion to the base of a pier. Since about 1967 special

FIG. 5.4. Serious abrasion by sand and shingle to concrete pier (courtesy: Cement and Concrete Association).

attention has been paid to the problem of the abrasion of concrete in marine structures by the Sea Action Committee of the Institution of Civil Engineers.

Investigations carried out by a member of the staff of the Cement and Concrete Association at the request of this Committee have shown that abrasion is a serious problem in a number of coastal areas. Little seems to be known with any degree of certainty about the factors which are involved in this type of abrasion of concrete. Therefore repairs to concrete damaged in this way have to be carried out on a more or less ad hoc basis. This means that the repaired concrete may not resist the abrasive effects of sea water, sand and shingle any better than the original concrete.

Arrangements were made in places where abrasion had occurred, to set up test panels made with a range of mix proportions, aggregates, and cements. The mix proportions appear to have been in the range of 1:6 and 1:7; the cements included ordinary, rapid hardening and sulphate-resisting Portland cements, and HAC. The aggregates used were all from 'natural sources' to BS 882. Information on the w/c ratios used does not appear to be available. The mixes used were, in the opinion of the author, rather lean for marine work; with a 1:6 mix and assuming a w/c of 0·55, the cement content would be about 310 kg/m^3 of concrete. With a 1:7 mix and a w/c ratio of 0·6, the cement content drops to about 280 kg/m^3. This can be compared with the recommendations in CP 110 for conditions of severe exposure of about 360 kg/m^3 and those given in this book for marine repair work of 400 kg/m^3.

Where plain mass concrete is used for marine work, then a cement content of 400 kg/m^3 is likely to bring problems of thermal contraction cracking. In such cases it may be desirable to design for a leaner concrete core with a cement rich, low w/c ratio mix for the exposed surface. This can be done by using a slip shutter between the two types of concrete which should be cast simultaneously.

In considering abrasion of marine structures, the author feels that reference can usefully be made to the information available on the abrasion of concrete floors of industrial buildings. It is admitted that environmental conditions are quite different, but a finish which will stand up to steel wheeled trolleys may give good results for wave-driven sand and shingle.

It is known from experience with industrial floors that concrete with a high cement content and low w/c ratio has considerably higher abrasion resistance than a leaner concrete using the same aggregate. It has been found cheaper to increase the cement content, lower the w/c ratio and use a plasticiser, than to import a harder aggregate from a long distance. The effect of the hardness of the aggregate on the abrasion resistance of the concrete floor is less than that of the overall quality of the concrete.

An interesting example of the abrasive resistance of high quality cement-rich concrete is the apron to the promenade at Littlestone, which was reconstructed by the Kent River Authority between 1960 and 1966 using ragstone slabs set in fine concrete. The concrete had mix proportions of 1 part sulphate-resisting Portland cement to $2\frac{1}{2}$ parts of flint gravel, with 4% air entrainment and 2% calcium chloride added to speed-up the setting and hardening. The w/c is not known, but the cement content of the mix was about 620 kg/m^3. The concrete has stood up very well to abrasion, in fact better than the ragstone slabs. There was no reinforcement in the concrete.

Information so far available suggests that initial surface damage to concrete, such as is caused by the incoming tide or premature removal of formwork, can form focal points for early and fairly rapid erosion.

At the present time it is not possible to make specific recommendations based on research or controlled experiments. Therefore when dealing with repair work it is necessary to base the proposals on experience gained in similar work and to adapt techniques which have been found satisfactory in other types of work, as far as this is practical.

The author therefore suggests the following:

(a) The concrete mix to contain a minimum of 400 kg cement/m^3 concrete.
(b) For Portland cements the w/c ratio should not exceed 0·45, and for HAC 0·40.
(c) The aggregate should be as abrasive resistant as possible; flint gravel and crushed granite are generally more resistant to abrasion than limestones and sandstones. If relatively small quantities of concrete are required, then it may be worthwhile considering the use of HAC with a special aggregate such as Alag, which is crushed and graded HAC clinker. Alag should only be used with HAC.
(d) Special precautions should be taken to protect the concrete (or gunite if this is used for the repair) from being damaged, even superficially, by tidal action.

For engineers who are interested in these problems and have the opportunity to experiment, the author suggests the following:

(1) An abrasion-resistant finish (in terms of industrial floors), can be provided by a metallic sprinkle finish on plastic insitu concrete. The concrete should be of marine quality as previously described. The sprinkle finish consists of finely divided iron premixed with cement; the thickness is usually about 3 mm, and the amount of iron is such as to give a weight of about 5 kg/m^2 of surface. For very heavy duty, the weight of iron is considerably increased, to about 45 kg/m^2, and surfacing is then laid as a topping about 10 mm thick, monolithic with the base concrete. This type of finish is only suitable for horizontal, or sloping surfaces with a gradient up to about 45°. Precast units can be made in this way, and then they can be used for vertical and steeply sloping members.

(2) Precast concrete slabs or blocks can be made with HAC and Alag aggregate. The precast units can then be set in a fine concrete made with either HAC and Alag aggregate, or HAC and a natural aggregate, or with a Portland cement and a natural aggregate. The concrete mix for both the precast units and the insitu jointing should be of marine quality.

(3) High quality concrete (as previously specified), can be used with the addition of about 3% by weight of random wire fibres. The function of the fibres is to improve impact resistance and increase resistance to cracking. The use of fibres in concrete is under development in various countries.

5.5 MARINE GROWTHS ON CONCRETE

Even in cold and temperate seas, marine growths appear very rapidly on concrete structures. In warm tropical waters, this growth is extremely rapid. In both cases, it is very difficult to remove the seaweed, barnacles etc. Questions are sometimes asked as to the long term effect of these growths on the durability of concrete. The author has found no published information which suggests that the durability of good quality concrete is adversely affected by seaweed, barnacles and the many thousands of other marine growths. There is probably a small removal of lime from the surface layers, but that is all; with adequate cover to the reinforcement of dense impermeable concrete, the effect is insignificant.

It is often necessary to remove these growths from slipways and steps, promenades and aprons to sea walls, because they make the surface of the concrete very slippery. In such cases, high velocity water jets are likely to give the best results.

The use of poisons mixed into the concrete or applied to the surface is dangerous in locations where people will walk or otherwise come into contact with the concrete. These antifouling preparations only inhibit the growth of marine organisms for a short period and therefore require regular renewal.

However, marine growths can seriously interfere with the carrying capacity of sea water intakes to power stations and research has been carried out to find some permanent technique for preventing these growths. A detailed report on this work has been published by the Civil Engineering Laboratory at the Naval Construction Battalion Centre in California; details of this report are given in the Bibliography at the end of this chapter. It was found that a durable antifouling concrete could be made with a porous expanded shale aggregate which had been impregnated with creosote and certain other toxic chemicals. The resulting concrete remained essentially free of marine growths for a period of four years. This special concrete was of medium strength, about 25 N/mm^2. The samples were exposed to marine conditions near the surface of the ocean and at a depth of about 35 m, off the coast of Cuba and California.

5.6 THE USE OF HIGH VELOCITY WATER JETS FOR CLEANING AND PREPARING CONCRETE SURFACES PRIOR TO REPAIR, AND FOR CUTTING CONCRETE

5.6.1 Cleaning and Preparing Surfaces

The use of high velocity (high pressure) water jets for a variety of mechanical engineering purposes, particularly in the marine field, has been in existence for the last 15 to 20 years. However, apart from the cleaning of structural steelwork, only in the past few years has this technique been used in the construction industry.

Recent work by the Construction Industry Research and Information Association (CIRIA) has shown that steelwork can be cleaned at least as effectively by high velocity water jets as by the normal methods of shot and grit blasting. Work in this field is still proceeding. The reason for the improvement in cleaning appears to be that the high velocity water removes all, or nearly all, traces of aggressive chemicals from the surface in addition to rust, scale etc. Shot and grit blasting does not remove the molecules of the aggressive chemicals which may then form the nucleus for fresh attack on the steel after the new coatings have been applied.

It is a logical step to extend the field of application to concrete. The technique can be used for cutting concrete, removal of laitance and exposure of the coarse aggregate, removal of deteriorated concrete and removal of marine growths, deposits and old coatings. The move into the concrete field has been slow, due to a number of factors including the natural inertia of the construction industry to accept new ideas, the lack of dissemination of information by the suppliers of the equipment, and the cost of the equipment (pump, motor, jetting gun, extension lances, control valves etc.).

The author has seen high velocity water equipment used for cutting holes in concrete and for producing an exposed aggregate surface suitable for receiving rendering and screed. This was carried out quickly and effectively, without vibration, noise and dust such as would be caused by the use of percussion tools, and grit blasting. For work on concrete, the pressure required to produce the necessary nozzle velocity would be about 400 atm or 6000 lbf/in^2.

The quantity of water used is quite small; for the type of work dealt with in this book it is likely to be about 55 litres/min (12 gpm) of which about 30 % is lost as mist and spray. This means that the water disposal problem in terms of quantity is a small one. The use of the high pressures mentioned above often gives rise to some disquiet on the part of prospective users. Water is relatively incompressible and the instant the water leaves the nozzle it is virtually unrestrained and assumes a velocity characteristic with practically no expansion. The great advantage of the use of high velocity water is that vibration, noise and dust are eliminated and the surface of the concrete is left clean and damp, which is the optimum condition for achieving good bond with subsequent layers of concrete or mortar.

5.6.2 Cutting Concrete

During repairs and alterations to concrete structures, it is often necessary to cut holes and chases through concrete walls and foundations. This work can be carried out by a variety of equipment, ranging from simple percussion tools and drills, thermic lance, diamond drilling (stitch drilling), and high velocity water jets.

The author's experience is that provided the presence of water can be accepted, high velocity water offers the best chance of a quick clean job. The water will not cut the reinforcement and therefore this has to be cut separately. The use of these jets has been described in the previous section.

However, there are occasions when the reinforcement must remain intact and therefore thermic lance and diamond drilling is not suitable. An example of this is where a new concrete wall has to be joined and bonded to

FIG. 5.5. High velocity water jet cutting slot in R.C. wall (courtesy: F. A. Hughes and Co. Ltd).

FIG. 5.6. Detail view of slot cut by high velocity water jet (courtesy: F. A. Hughes and Co. Ltd).

an existing one. Figure 5.5 shows high velocity water jets being used for this purpose, and Figure 5.6 shows the concrete slot completed and ready to receive the new concrete. The wall was approximately 225–300 mm thick and 6·0 m high; the slot was tapered from 300 mm on the outer face to 50 mm at the rear. The work was completed in $7\frac{1}{2}$ h.

When it is necessary to cut through reinforcement and the work can be carried out above low water level, the thermic lance can be used. This piece of equipment works on the principle that when carbon steels are heated at about 900 °C (which is a bright red colour), they will burn in an oxygen-rich atmosphere; the flame arising from this combustion has a temperature of about 3500 °C. This temperature is high enough to melt concrete, clay bricks and steel. The thermic lance can be used for a variety of work such as cutting holes in concrete, cutting up concrete sections and cutting openings in concrete walls. Under no circumstances should it be used where there is any risk of fire or explosion. For further details of this technique the reader is referred to the Bibliography at the end of this chapter.

5.7 CATHODIC PROTECTION OF REINFORCED CONCRETE STRUCTURES

Sea water is very aggressive to ferrous metals. While high quality dense impermeable concrete of adequate thickness will give complete protection to the reinforcement this protection will only last as long as the integrity of the concrete cover remains unimpaired. Small areas of relatively porous (under-compacted) concrete, reduction in depth of cover to the steel, thermal contraction cracks, physical damage, can all result in sea water eventually getting through to the reinforcement. Once corrosion starts it becomes progressively worse as the corrosion products (rust) spall and crack the concrete thus admitting more sea water to the steel.

Sooner or later all marine structures require repair and this is always a costly business. The corrosion of steel by sea water is an electro-chemical process and therefore it is logical (even though it may be difficult in practice) to protect the steel by electrical means, *i.e.* by cathodic protection. It is common practice to use cathodic protection on steel structures, such as jetties, as well as on steel underground pipelines. This type of protection is in addition to standard coating procedure.

It is easier to protect reinforcement cathodically in a concrete structure immersed in sea water than reinforcement in a land-based structure because the sea water forms an electrolyte. A number of reinforced concrete marine structures have been successfully provided with cathodic protection. From published reports and papers it is clear that greater attention is now being given to this technique than in the past. There are however many difficulties and many problems have to be solved before cathodic protection to marine structures can be acknowledged as a practical method of protecting the reinforcement.

Reinforced concrete is now being used for oil production platforms in the sea at far greater depths than was thought possible only a few years ago. At the present time no one knows for sure how these vast structures will be maintained and when necessary, repaired. It is unlikely that the concrete itself will deteriorate; the material to cause the trouble will be the steel reinforcement. The author feels that greater use can be made of cathodic protection, but this will involve research and development.

FIG. 5.7. Modern design for R.C. dock gates. View prior to casting and installation of cathodic protection (courtesy: Laing Offshore Ltd and P.I. Corrosion Engineers Ltd).

Figure 5.7 shows insitu reinforced concrete dock gates in which the reinforcement has been cathodically protected. This is believed to be the largest reinforced concrete structure which has been cathodically protected in the UK.

Selected references on cathodic protection are included in the Bibliography at the end of this chapter. However, published papers generally refer to theoretical and laboratory studies and small scale trials rather than site application. The principal problems relate to the difficulty of bonding the reinforcement to secure electrical continuity and ensuring as far as practical that the 'path' for the electric current from the cathode in the steel reinforcement to the anode and back is not too 'difficult'. In other words the resistance must not be too high. Dry concrete will have a much higher resistance than wet concrete.

5.7.1 General Principles of Cathodic Protection

In above-ground steel structures the usual method of protecting steel against corrosion is by means of coatings. These may be of another metal, as in galvanising or metal spraying, or various paint systems, as well as the more recently introduced use of plastic films.

In Chapter 2 it was stated that the corrosion of metals in a conducting environment is electro-chemical in nature, and this is due to all metals attempting to revert to their natural ore state. Exposed steel in a moist environment will corrode due to differences in electrical potential on the surface of the metal itself. These areas form anodes and cathodes, which, by definition, permit a flow of electric current from anode to cathode. It should be noted that electron flow, which is by inference a part of the same process, is in the reverse direction. The multitude and extent of the consequent electro-chemical reactions result in the generation of a continuous electric current made up of a multiple of small anodic and cathodic areas. With steel and some other alloy structures buried underground or immersed in water (*e.g.* marine structures) corrosion control is often exercised by electrical methods and known as 'cathodic protection'. This is usually achieved by the application of a continuous direct current from an external source. It is not possible, nor desirable in this book, to enter into the detailed theory and practice of cathodic protection, so only brief mention will be made of the principles involved.

When two dissimilar metals are joined together in an electrolyte a current is produced as a function of the electro-chemical series of metals. For example, if steel and zinc are connected, a current will flow from the zinc to the steel because the zinc is anodic to steel. This reaction takes place only if there is an electrolyte present between the two dissimilar metals, although it does not necessarily have to be water, inasmuch as severe corrosion can take place in soils ranging from wet clay to rock. With the zinc/steel combination just mentioned, the zinc will corrode and the steel will not, and the overall effect is for the steel to have become inhibited against corrosion attack by virtue of the anode material sacrificing itself to afford protection to the steel. Hence the term 'sacrificial anode protection'.

Alternatively, the same result can be obtained by the introduction of a controlled d.c. electricity application to the structure from either an a.c. supply source, diesel generators, thermo-electric equipment, solar cells or even high efficiency windmills. The structure is connected to the negative supply (or cathode) and the positive to an introduced anode which is chosen to have semi-inert (or non-corrodable) properties. This is known as the 'impressed current' method.

Thus there are two practical ways of introducing cathodic protection which can be summed up as:

(a) by connecting the steel to a metal which is less noble in the electrochemical series. The following list shows relative activity of metals which varies from the highly reactive alkaline metals such as potassium and sodium to noble metals like gold. The list indicates a basic series in which the metals high on the list will become anodic to those lower down and hence provide protection. This is the basis of the principle previously described as sacrificial anode protection, and depends on one metal being designed to corrode and so prevent another corroding.

Sodium
Magnesium
Zinc
Galvanised iron
Aluminium
Mild steel
Cast iron
Stainless steel (Cr based)
Lead
Bronze
Brass
Copper
Stainless steel (passive)
Silver
Gold

Note the relative positions of commonly used metals or alloys, *e.g.* galvanised iron/steel/brass/copper/stainless steel, which is the basis of most forms of introduced electrolyte corrosion damage.

(b) by the application of an external current of sufficient intensity to 'swamp' the corrosion current. The impressed current is d.c. with the positive side of the output connected to a purpose made anode, and the negative to the structure being protected. The anode can, in theory, be of any material which will conduct electricity, but in practice the range of suitable materials is limited principally to ensure system longevity. Anodes in this system are comparatively inert and are designed to last much longer than sacrificial anodes. When correctly selected and used in a designed system, such anodes must have a useful life of 20 years and more. The materials available are based on graphite, high silicon-iron, platinised titanium, tantalum or niobium or the lead/silver alloys which are principally used in marine installations. Although platinum and its alloys appear expensive they are becoming more popular because of their electrochemical properties and light weight.

In considering the application of cathodic protection to a structure, the following questions arise and must be satisfactorily answered:

(i) Surface area to be covered and its treatment, *e.g.* bare, old paint, newly applied high grade coating
(ii) The environment surrounding the structure, *e.g.* wet conditions, pH, chemical characteristics
(iii) What type of protection might be the most suitable in the given circumstances, *i.e.* sacrificial anodes or impressed current
(iv) The general circuit configuration, including the number, location and type of anodes to be used
(v) Is there any likelihood of the protective current flowing outside the network which is supposedly being protected? If this risk is likely to arise, then anode locations must be changed or interference testing contemplated.

The assessment of corrosiveness of the environment is in inverse proportion to the conductivity; the most practical way of assessing any form of electrolyte, whether liquid or solid, is by testing its electrical resistivity utilising special megger type equipment. The following table indicates in a very general way the relationship between the environment and its corrosiveness in terms of resistivity:

TABLE 5.1
(*from B.S.I. CP* 1021: 1973: *Cathodic Protection*)

Resistivity (ohm–metres)	Degree of corrosiveness
Up to 10 ohm/m	Severely corrosive
10 ohm/m–100 ohm/m	Moderately corrosive
100 ohm/m and above	Slightly corrosive

It will be seen that virtually no environment can be considered as non-corrosive, because every metal has a tendency to revert back into the basic form from which it was extracted as an ore. The environment controls the speed of this reaction, and ranges between highly corrosive conditions, such as found in sea water at 20 ohm/cm, to virtually non-corrosive conditions in granite rock at 500 000 ohm/cm.

It is generally desirable that buried or immersed steelwork should, in addition to cathodic protection, be complemented with suitable coatings or wrappings. This additional protection is provided in order to reduce the rate of dissolution of the sacrificial anodes, and so prolong their useful life, or if impressed current is utilised, to reduce the power consumption.

There are many reported cases of the use of cathodic protection to reduce or eliminate corrosion of the prestressing wires in prestressed concrete water mains. The work of Unz and Spector in Israel, Baeckmann in Germany and Franquin in France are well known.

BIBLIOGRAPHY

Kalousek, G. L., Porter, L. C. and Benton, E. J., *Cem. Conc. Res.*, 1972 (2), pp. 79–89

Halstead, P. E., Behaviour of structures subjected to aggressive waters. Paper at Inter-Association Colloquium on *Behaviour in Service of Concrete Structures*, Liege, June, 1975, p. 11

Building Research Establishment, *Concrete in Sulphate-bearing Soils and Ground Waters*. Digest 174, 1975, HMSO, London, p. 4

American Concrete Institute, Guide for the protection of concrete against chemical attack by means of coatings and other corrosion-resistant materials. Report by ACI Committee 515, *Journal A.C.I.*, 1966, Dec. 1305–1390

Lea, F. M. and Desch, C. H., *The Chemistry of Cement and Concrete*, 2nd ed. (Revised). Edward Arnold Ltd., 1970, London, p. 637

American Concrete Institute, *Manual of Concrete Practice*. Part 3—Products and Processes. 1972, A.C.I., Detroit, pp. 515–23 to 515–73

Shacklock, B. W., *Concrete Constituents and Mix Proportions*, Cement and Concrete Association, London, 1974, p. 102

Murdock, L. J. and Blackledge, G. F., *Concrete Materials and Practice*. Edward Arnold Ltd, London 1968, 4th ed. p. 398

Building Research Establishment, *The Durability of Reinforced Concrete in Sea-Waters*. Twentieth Report of the Sea Action Committee of the Institution of Civil Engineers, National Building Studies, Research Paper No. 30. 1960, HMSO, p. 42

Houston, J. T., Atimtay, T. and Ferguston, P. M., *Corrosion of Reinforcing Steel Embedded in Structural Concrete*. Research Report 112–1F, Centre for Highway Research, University of Texas, Austin, U.S., March 1972, p. 131

Perkins, P. H., *Floors—Construction and Finishes*, Cement and Concrete Association, London, 1973, p. 132

American Concrete Institute, *Manual of Concrete Practice* Part 1—*Erosion resistance of concrete in hydraulic structures*, 1968, A.C.I., Detroit, pp. 210–1 to 210–13

Kenn, M. J., Factors influencing the erosion of concrete by cavitation, *CIRIA Technical Note*, (1) London, July 1968, p. 15

Vickers, A. J., Francis, J. R. D. and Grant, A. W., Erosion of sewers and drains. *CIRIA Research Report* (14) London, October 1968. p. 20

Biczok, I., *Concrete Corrosion and Concrete Protection*, 4th ed. Hungarian Academy of Sciences, 1967 p. 543

Hausman, B. A., Criterion for cathodic protection of steel in concrete. *Materials Protection*, **8**(10) October (1969) pp. 23–25

Everett, L. H. and Treadaway, K. W. J., *The Use of Galvanized Steel Reinforcement in Building*. Building Research Station Current Paper CP3/70 January 1970, p. 10

TREADAWAY, K. W. J. AND RUSSELL, A. D., *Inhibition of the Corrosion of Steel in Concrete*. Building Research Station Current Paper CP 82/68, December 1969, p. 5

BRYSON, J. O. AND MATHEY, R. C., Surface condition effect on bond strength of steel beams embedded in concrete. *Proc. A.C.I.* **59**(3) (1962), pp. 397–406

COOK, A. M., A comparison of the bond strength of black and galvanized plain reinforcing bars to concrete. A confidential report, not published. The Cement and Concrete Association of Australia, TM95, April 1974, p. 16

THE CONCRETE SOCIETY, *Underwater Concreting*. Technical Report 52.018. Cement and Concrete Association, 1971. p. 12

PURDEY, P. H., Notes on marine applications of jet cutting. Paper No. D2. *First International Symposium on Jet Cutting Technology*, Coventry, April 1972, pp. 13–23

PURDEY, P. H. Field use of high velocity water jets and the contribution to safety and training. Paper No.F2, *Second International Symposium on Jet Cutting Technology*, Cambridge, April 1974, pp. 11–18

MUSANNIF, A. A. B., Cutting concrete down to size, *Civil Engineering*, June 1975, pp. 37 & 39

AKAM, E. A., *Demolition*. Building Research Establishment Current Paper CP 12/72, 1972

H.M. GOVERNMENT, *The Factories Act*, 1961, HMSO, London

H.M. GOVERNMENT, *The Health and Safety at Work Act*, 1974, HMSO, London

NATIONAL ASSOCIATION OF CORROSION ENGINEERS, *Cathodic Protection*. Report of Committee July 1951, p. 33

HEUZÉ, B., Cathodic protection of steel in prestressed concrete. *Materials Protection*, November 1965, 57–62

SPECTOR, D., A study of failures and cathodic protection of reinforced concrete pipes. *Aqua* (4) (1964) pp. 7–24

UNZ, M., Cathodic protection of prestressed concrete pipes. *Corrosion*, June 1960, 289t–297t

UNZ, M., Interpretation of surface potentials in corrosion tests—in particular, steel embedded in concrete, *Israel Journal of Technology*, **4**, (1966), pp. 243–255

CORNET, I., Steel, concrete and salt water. Paper published by the University of California, 1974, pp. 215–225

MURAOKA, J. S. AND VIND, H. P., *Anti-Fouling Marine Concrete*. Civil Engineering Laboratory, Naval Construction Battalion Centre, Cal., USA, May 1975, p. 22

GJORV, O. E., Long-time durability of concrete in seawater. *Journal A.C.I.*, Jan. 1971, pp. 60–67

CHAPTER 6

Protection of Concrete Structures Against Chemical Attack

In previous chapters reference has been made to 'extreme' and 'severe' conditions of exposure and 'exposure to an aggressive environment'. While these terms are often used to indicate the same sort of environmental conditions, the author prefers to consider 'extreme' and 'severe' as describing natural physical conditions of wind, rain, freeze and thaw, and the 'aggressive' environment as referring to conditions involving chemical attack on the structure or member. The vast majority of environmental conditions do not require any protective measures to be adopted, but there are circumstances where it is obvious that the concrete must be completely separated from the aggressive chemicals if severe deterioration is to be prevented. However, it is the 'grey' areas in between these two extremes which are usually the most difficult to deal with in an economic and practical way.

There is obviously a difference in approach in preparing the design and specification for a new structure when it is anticipated that chemical attack on the concrete will occur and the steps which have to be taken when dealing with an existing structure which has suffered chemical attack. The former (new structures), may at first sight appear to be quite straightforward, but in practice, this is not necessarily so. It is frequently difficult to assess with any degree of accuracy the likely severity of attack. An example is a sewer or sewage tank containing trade effluent; the actual chemicals which will be present, their concentration and temperature, will all have a profound influence on the selection of the protective measures required.

The first problem is to decide whether any special protective measures in the form of barrier layers are required at all, or whether good quality concrete made with ordinary Portland, sulphate-resisting Portland, or HAC, will give the necessary life, either on their own or as an additional thickness to form a sacrificial layer.

When dealing with deterioration of an existing structure it can be difficult to determine exactly what has caused the attack unless the aggressive agents are still present. It is most important to obtain the services of a chemist experienced in concrete technology, otherwise wrong conclusions can be drawn from the chemical analysis of the damaged concrete.

When it has been decided that the concrete will need a protective barrier the solution of the problem then lies in the selection of the barrier layer and the best method of application. With layers which are bonded to the base concrete, as is usually the case, the correct preparation of the surface of the concrete is of vital importance.

When selecting a coating or lining it is necessary to know as accurately as possible, the physical and chemical conditions under which the material will have to operate. There is no essential difference in a coating and a lining, although the former is generally used to denote an external protective barrier and the latter an internal one. The following are the principal questions which must be answered before a decision can be taken on the protective material required:

(1) Of what material is the substrate to which the coating/lining will be applied, *i.e.* insitu reinforced or prestressed concrete, precast concrete units, gunite, cement/sand rendering and the background to which this has been applied?
(2) Is the coating/lining to have as long a life as possible, or can it be considered as temporary?
(3) Will there be any adverse physical conditions such as abrasion or impact?
(4) What will be the operating temperature range?
(5) What aggressive chemicals will the coating/lining have to resist?
(6) Are there any special site conditions likely to affect the application of the coating/lining, such as high or low ambient temperature, high or low relative humidity, availability or otherwise of skilled labour?
(7) Are there any overriding restrictions on cost?

It is important that detailed consideration be given to all the above questions, and then decisions can be taken in the light of all the answers. Such matters as the degree of flexibility required in the lining, its chemical resistance at operating temperature, the practical operating temperature range including shut-down periods, are all of vital importance.

Blue acid-resistant clay bricks complying with BS 3679: Acid Resisting Bricks and Tiles, bedded and jointed in an appropriate chemically resistant mortar, may fulfil all requirements except temperature movement in the structural shell. Differential movement between the shell and the lining may necessitate the use of a flexible membrane to separate the shell from the lining.

It is obviously impossible in a single chapter, or even in a complete book to cover adequately all types of structure and all possible corrosive chemicals and environmental conditions.

While this book as a whole is concerned with repairing existing structures, in this chapter some emphasis is placed on precautionary measures which can be adopted in the design stage. This is particularly important in the case of sewerage systems where sulphuric acid formed by two stage bacterial action has attacked the concrete. In these circumstances steps should be taken as far as this is practical, to reduce the aggressive conditions after the repairs have been carried out. If this is not done, trouble may start again.

6.1 AGGRESSIVE CHEMICALS AND PORTLAND CEMENT CONCRETE

The range of chemicals used in modern industry is staggeringly large and each case has to be dealt with as a separate problem and given careful study.

Portland cement concrete is vulnerable to attack from all types of acids, sulphates in solution, ammonium salts in solution, as well as a number of other chemicals. Brief notes on the more commonly used compounds are given below:

Acids The degree of attack depends mainly on the chemical composition of the acid, its concentration and temperature, whether it is continuously renewed, and on the impermeability and cement content of the mix and the type of cement and aggregate used.

Sulphates in solution Reference should be made to the detailed information on aggressive ground water given later in this chapter. The only additional comment is that with stored liquids, the temperature may be appreciably above ambient, and increase in temperature usually increases chemical activity.

Brine This is usually a concentrated solution of sodium chloride (common salt). At normal temperature, this is not aggressive to concrete, but strong calcium chloride brine is likely to cause some deterioration in concrete or mortar. It must also be realised that all brines are very aggressive to steel reinforcement and so maximum precautions must be taken to ensure impermeable and crack-free concrete and mortar.

Creosote This contains phenols and will slowly disintegrate concrete and mortar.

Lubricating oils If these contain fatty acids they are likely to attack concrete slowly; also they have high penetrating powers and if containing acids are potentially dangerous to reinforcement.

Petroleum oils These include petrol, paraffin and diesel oil. Generally these oils are not aggressive to concrete and mortar provided they have a low fatty acid and sulphur concentration. However, they have high penetrating characteristics, and are likely to seep slowly through concrete unless an inert barrier is provided. The seepage is very slow and will show as dark stains and moist oily patches on the outside of the container.

Sulphur (molten) Liquid sulphur is not aggressive to Portland cement products.

Caustic soda and other caustic alkalis Sodium hydroxide up to a concentration of about 10% is harmless to good quality concrete and mortar; concentrations above 10% are likely to cause slow disintegration. Increase in concentration and temperature of the solution will usually increase the speed of attack with most caustic alkalis.

Arsenic compounds It is rare for these to come into contact with concrete, but two authorities on the subject (Imre Biczok and A. Kleinlogel) state that arsenous compounds are unlikely to attack good quality concrete at normal temperatures.

Ammonium and magnesium compounds Certain compounds of ammonium and magnesium are aggressive to Portland cement concrete by what is known as 'cationic' attack. This is the replacement of the lime in the concrete (principally the cement) by other cations. This is effected by magnesium and ammonium ions and the result is a serious weakening of the concrete.

Urea Urea is sometimes used as a deicing compound and will attack concrete. The exact mechanism of the attack is not fully understood.

Copperas This is a trade name for ferrous sulphate. When dissolved in water it gives an acid reaction to litmus. The solution will attack Portland cement concrete.

Calcium hypochlorite (bleaching powder) A concentrated solution of this compound may attack concrete slowly. Calcium hypochlorite is alkaline in reaction and in the concentrations used for sterilisation of water, no deterioration of good quality concrete will occur.

Sugar Sugar will slowly attack even good quality Portland cement concrete.

Milk Fresh milk is not aggressive to concrete, but sour milk and milk products are, due to the formation of lactic acid.

Fruit and vegetables The juices derived from fruit and vegetables contain sugar and in most cases organic acids as well, and will therefore attack concrete slowly.

Wine Wine will not attack concrete, but to prevent contamination of the wine by alkalis from the concrete, wine tanks should be lined.

Sewage Domestic sewage is not aggressive to Portland cement products; septic sewage, with a pH below 6·5 may attack concrete and mortar. Sewage containing trade wastes should always be carefully investigated. Sewers and sewerage structures are dealt with in a separate section in this chapter.

Sulphuretted hydrogen or hydrogen sulphide (H_2S) Sulphuretted hydrogen (also known as hydrogen sulphide) will not attack good quality concrete. When dissolved in water it forms hydrosulphuric acid which is one of the weakest acids known and would have little or no erosive effect on good quality concrete. However under certain conditions it can be converted to sulphuric acid by aerobic bacteria and this acid is very aggressive to both

Portland cement and HAC concretes. This is dealt with in some detail later in this chapter.

Sea water The effect of sea water on Portland cement concrete and mortar is discussed in detail in Chapter 5 which deals with repair and protection of marine structures.

Distilled water Hot distillate (*e.g.* from sea water distillation plants) is very aggressive to concrete.

6.2 STRUCTURES IN AGGRESSIVE SOIL AND GROUND WATER

The first and most important point is that chemicals in the dry state will not attack concrete; this applies particularly to sulphates as these are often found in the form of lenses in clay. Table 6.2 shows that the tolerable level of sulphate concentration in relatively dry well-drained soils is about four to six times the acceptable limit when the sulphates are in solution in ground water. In most soils it is the concentration of aggressive chemicals in the ground water which decides whether or not protective measures have to be taken.

Whenever there is any doubt about the possible attack on the concrete, a full chemical analysis of the soil and ground water is recommended. Such analyses require considerable experience for realistic interpretation.

There are two main categories of ground water, namely, naturally occurring ground water, and ground water from industrial tips. Both types can be aggressive to concrete, but that from industrial tips is likely to be far more dangerous and difficult to deal with.

6.2.1 Acids in Ground Water

When the pH of the sub-soil and/or ground water is below the neutral point of 7·0 the water is acidic and is liable to attack concrete made with any type of Portland cement. The severity of the attack depends on a number of factors, the principal ones being:

- (i) the type and quality of acid present
- (ii) whether the acids are likely to be continuously renewed
- (iii) the pressure and velocity of flow of ground water against the concrete
- (iv) the cement content and impermeability of the concrete and the type of aggregate.

The pH by itself does not give any indication of the type or amount of acid present. It is a measure of the intensity of the acidity, and can be considered as analogous to voltage in electric circuits. Thus while pH measurements are useful as a rough guide, and clearly show the presence (or absence) of acids, they should, in most cases, be supplemented by chemical analysis to determine the nature and quantity of the acid present. In natural waters of low pH, the acidity is usually caused by organic acids and dissolved carbon dioxide; to this is sometimes added sulphurous and sulphuric acids derived from sulphur compounds in peaty sub-soils.

Some research has been carried out on the effect of water from upland gathering grounds on Portland cement concrete. There is evidence which suggests that dissolved carbon dioxide is more aggressive to concrete than the organic acids normally found in these waters. The carbon dioxide causes leaching of lime from the concrete. In cases of acid attack on concrete the use of a limestone aggregate can be beneficial because the acidity in the water which is in contact with the concrete is neutralised by both the cement paste and the aggregate. If an inert aggregate is used, such as a flint gravel, then the attack is concentrated on the cement paste alone and the damage is likely to be more severe.

An exception to this is where the chemical characteristics of the water render it liable to attack the limestone selectively, but such cases are rare. However, this possibility exists, and therefore the chemical analysis of the water should be interpreted by an experienced chemist.

The following table is intended as a guide only; each case should be treated as an individual problem. It is assumed that the concrete is high quality.

TABLE 6.1

Acids in naturally occurring ground water

(i) Significant attack probably unlikely	(ii) Slight attack probable	(iii) Appreciable attack probable	(iv) Severe attack probable
pH 7·0–6·5	pH 6·5–5·5	pH 5·5–4·5	pH below 4·5

It is recommended that the quality of the concrete be increased (higher cement content and lower w/c ratio), as the severity of attack rises, as detailed below.

The table should be read with the text and the notes which follow. Where the pH of the water is below 7·0 it is prudent to carry out a full chemical analysis unless there is previous experience which shows that the water does not attack the concrete. This should certainly be done where the pH is below 6·5. It is important that all the concrete used should be fully compacted and properly cured, and that reinforcement should have a minimum cover of 40 mm, which should be increased to 50 mm for conditions (ii) to (iv).

(i) Where conditions are such that attack on the concrete is considered unlikely, then ordinary Portland cement concrete containing not less than 300 kg cement/m^3, and with a w/c ratio not exceeding 0·5, should be used.

(ii) Where the pH is 6·5–5·5, the concrete can be made with ordinary Portland cement, but the cement content should not be less than 330 kg/m^3, and the w/c ratio should not exceed 0·50.

(iii) The cement content should be increased to 360 kg/m^3, and the w/c ratio reduced to 0·45. This low w/c ratio may require the use of a plasticiser to obtain the necessary workability for full compaction of the concrete.

(iv) Where severe attack is probable, concrete of the quality in (iii) above should be used. In addition, there should be either a substantial sacrificial thickness of the same high quality concrete, or the concrete should be protected by a coating of inert material. The type of coating and its thickness will require careful selection, taking into account the aggressive chemicals present and site conditions. Generally the choice will lie between insitu materials and preformed sheeting. The insitu coatings include such materials as epoxides, polyurethanes, polyesters and bituminous compounds. The minimum thickness is about 0·50 mm and this should be attained in not less than three coats so as to avoid pinholes. Mastic asphalt can also be used and details of its application are given later in this chapter. Preformed sheeting can be butyl rubber, polyisobutylene, PVC, and polyethylene. It is advisable for the sheeting to be not less than 0·75 mm thick and all joints should be hot or solvent welded. It is important that coatings and sheetings be adequately protected against damage by back-filling; the actual protection provided will depend on the vulnerability of the material used to abrasion and impact. A further point is that bituminous and asphaltic materials do not bond well to damp concrete.

Sulphate-resisting Portland cement does not differ significantly from ordinary Portland in its acid-resisting qualities. However, it would be advantageous to use it if the acidic ground water contained sulphate ions (sulphates in solution).

HAC is generally more resistant to dilute acids and sulphates than Portland cement. However, advice should be obtained from the manufacturers in each particular case.

Under most conditions found in practice involving naturally occurring ground water (and provided mineral acids are not present), high quality dense concrete provides adequate durability. This can be augmented by cut-off drainage and back-filling with calcareous material to help neutralise acidity. An additional thickness of concrete to act as a sacrificial layer is often a practical solution. Additional protection, if such is considered necessary, can be provided by the provision of 1000 gauge polyethylene sheets with a rubber-bitumen adhesive on one side (trade name 'Bitu-Thene'). These should be laid below the concrete and carried up the sides to

TABLE 6.2

Requirements for concrete exposed to sulphate attack

Class	Concentration of sulphates expressed as SO_3: In soil: Total SO_3	In soil: SO_3 in 2:1 water: soil extract	In groundwater	Type of cement	Requirements for dense, fully compacted concrete made with aggregates meeting the requirements of BS 882 or BS 1047: Minimum cement content (kg/m^3), Nominal maximum size of aggregate 40 mm	20 mm	10 mm	Maximum free w/c ratio
1	< 0·2% 2 000 ppm	—	< 300 ppm	Ordinary Portland or Portland blastfurnace	240	280	330	0·55
2	0·2–0·5% 2 000–5 000 ppm	—	300–1 200 ppm	Ordinary Portland or Portland blastfurnace	290	330	380	0·50
				Sulphate-resisting Portland	240	280	330	0·55
				Supersulphated	270	310	360	0·50
3	0·5–1·0% 5 000–10 000 ppm	1·9–3·1 g/l	1 200–2 500 ppm	Sulphate-resisting Portland or super-sulphated	290	330	380	0·50
4	1·0–2·0% 10 000–20 000 ppm	3·1–5·6 g/l	2 500–5 000 ppm	Sulphate-resisting Portland or super-sulphated	330	370	420	0·45
5	Over 2% Over 20 000 ppm	> 5·6 g/l	> 5 000 ppm	As for Class 4, but with the addition of adequate protective coatings of inert material such as asphalt or bituminous emulsions reinforced with fibreglass membranes				

Notes: This table applies only to concrete made with aggregates complying with the requirements of BS 882 or BS 1047 placed in near-neutral ground waters of pH 6–9, containing naturally occurring sulphates but not contaminants such as ammonium salts. Concrete prepared from ordinary Portland cement would not be recommended in acidic conditions (pH 6 or less); sulphate-resisting Portland cement is slightly more acid-resistant but no experience of large-scale use in these conditions is currently available. Supersulphated cement has given an acceptable life, provided that the concrete is dense and prepared with a free w/c ratio of 0·40 or less, in mineral acids down to pH 3·5.

The cement contents given in Class 2 are the minima recommended by the manufacturers. For SO_3 contents near the upper limit of Class 2, cement contents above these minima are advised.

For severe conditions, *e.g.* thin sections, sections under hydrostatic pressure on one side only and sections partly immersed, consideration should be given to a further reduction of w/c ratio and, if necessary, an increase in cement content to ensure the degree of workability needed for full compaction and thus minimum permeability.

The above table and the notes are based on B.R.S. Digest No. 174, February 1975, and are reproduced by courtesy of the Building Research Station.

The author considers that the following additional notes are valid:

(a) When the pH of the ground water and/or soil is below 6·5, a full chemical analysis is recommended, and decisions should be taken on such analysis and not on the pH alone.
(b) For class 4 soils and ground waters, the use of correctly proportioned and mixed HAC concrete could be considered. Before this cement is used, the manufacturers should be consulted. In the UK, special approval is now required under the Building Regulations.
(c) Supersulphated cement is no longer made in the UK.
(d) For protection of concrete in class 5 soils and ground waters, there are now other materials on the market than asphalt and bitumen which are likely to be equally effective as a barrier, and will bond strongly to damp concrete.
(e) The statement in the B.R.S. notes that: ‘Concrete prepared from ordinary Portland cement would not be recommended in acidic conditions (pH 6 or less),’ is open to question. The author’s comments and recommendations for concrete in acidic ground water given in the preceding section should be studied.

above ground water level, and precautions must be taken to prevent damage to this barrier layer.

6.2.2 Sulphates in Solution in Ground Water

Table 6.2 shows a range of ground conditions containing sulphates in solution in various concentrations and recommendations for type of concrete to be used. When the table is used to help solve practical problems arising from sulphate concentrations, the author recommends that the following should also be noted:

(i) The sulphates most commonly found occurring naturally in ground water are: calcium sulphate (gypsum) ($CaSO_4$); magnesium sulphate (Epsom salts) ($MgSO_4$); sodium sulphate (Glauber's salts) (Na_2SO_4).

(ii) Calcium sulphate is less soluble at normal temperatures than the other sulphates mentioned, and the solution is saturated at about 2000 mg/l (ppm), which corresponds to about 1200 mg/l of sulphur trioxide (SO_3). This means that under normal conditions of temperature, the concentration of calcium sulphate, expressed as sulphur trioxide, cannot exceed about 1200 mg/l.

(iii) Sodium and magnesium sulphates are very soluble in water and therefore the concentration can be much higher than with $CaSO_4$.

(iv) Magnesium sulphate is more aggressive to Portland cement concrete than the sulphates of sodium, calcium and potassium, when in equal concentrations, and this should be taken into account when deciding on the precautions to be taken. When the concentration is near the top of a range, the recommendations for the next higher class should be adopted.

(v) Ammonium sulphate has not been mentioned as it is unusual to find this in natural ground water and soils, but it is quite usual to find this and other ammonium compounds in industrial tips. Generally, ammonium salts are very aggressive to Portland cement concrete, and special precautions may have to be taken for which expert advice should be obtained.

(vi) The first column of Table 6.2 refers to 'relatively dry, well drained soils'. As previously stated, concrete is not attacked by sulphates in the dry state, and therefore it is assumed that the sulphates in this type of ground will in fact go into solution slowly. This is why the tolerable concentration in the soil is four to six times that in the ground water.

(vii) The sulphate concentration is shown as the equivalent of sulphur trioxide (SO_3). Sometimes sub-soil investigation reports show this as SO_4; conversion from SO_4 to SO_3 can be made as follows:

$$SO_3 = SO_4 \times \frac{80}{96} = 0{\cdot}83\, SO_4$$

Table 6.2 shows that the first line of defence against sulphate attack is a high quality, impermeable concrete; the second line is the use of a Portland cement with a low tricalcium aluminate (C_3A) content. In sulphate-resisting Portland cement to BS 4027, the C_3A content is limited to a maximum of 3·5%, while in ASTM specification C 180.68 (Type V cement), the C_3A content must not exceed 5%. Recent work by Kalousek and Porter suggests that the addition of pozzolanas to a concrete mix where the cement has a low C_3A content, will give additional protection in the long term against sulphate attack. However, at the time of writing this book, there has been no reliable published information to show that the addition of pozzolanic material such as selected pfa, to an ordinary Portland cement, will provide the same degree of sulphate resistance as sulphate-resisting Portland cement.

Figure 6.1 shows the effect of sulphate attack on concrete made with ordinary Portland cement compared with control cubes of similar concrete stored in normal tap water.

FIG. 6.1. The effect of sulphate solution on OPC concrete. (Left) Concrete in sulphate solution, note expansion. (Right) Concrete in ordinary tap water. (Courtesy: Cement and Concrete Association.)

6.2.3 INDUSTRIAL TIPS

The protection of concrete foundations in an industrial tip can present very difficult practical problems. The problems consist in deciding on a suitable protective system for the concrete and for its application to the concrete. It is usually not particularly difficult to find a material which will resist attack from the chemical compounds present in the tip, the real problem is how to apply it to the concrete. For example, how can protection be provided to the concrete in a bored pile or a precast driven pile? Even strip foundations and column bases present practical problems.

Obviously the first step is to arrange for an adequate number of trial bore holes; the number is likely to be considerably larger than on a site consisting of virgin ground. The next step is a careful examination of the borehole logs and chemical analyses of the subsoil and ground water; these should be interpreted in a realistic way so that their significance is clearly understood.

While there are a number of excellent books on the design and construction of foundations, and on soil mechanics, the author has found very little published information on how the problem of chemical attack on concrete foundations in an industrial tip can be dealt with. All that can be done here is to outline basic principles and set out various ideas in the hope that they will be of assistance to engineers faced with these serious problems.

It is obviously impossible to list the chemicals which will be found in an industrial tip and will be aggressive to concrete. However, experience suggests that in general, they are likely to include mineral acids, such as sulphuric, hydrochloric and nitric, ammonium compounds, phenols and sulphates. The sub-soil survey is certain to indicate that the aggressive materials are not uniformly distributed, but occur in certain areas at varying depths. The presence of ground water above the level of the bottom of the foundations will increase the aggressiveness of the chemicals and promote their distribution throughout the site.

A solution which may be both practical and economic on certain sites is to remove by bulk excavation, all the contaminated soil, and provide adequate sub-soil drainage and back-fill with non-aggressive materials. Limestone and chalk can be particularly useful in neutralising acidity as the ground water may have become contaminated over a fairly wide area.

If bulk excavation is not possible, then a well designed system of sub-soil drains should be provided and protection given to the concrete. The actual protection given will depend on the presence of ground water, the chemicals present, and the type of foundation (piles, raft, strip footings, column bases).

The use of HAC concrete is always worthy of consideration as it exhibits greater resistance to certain dilute acids and sulphates than does Portland cement. It is essential that the basic principles of mix design, placing and curing for HAC concrete should be carefully followed in order to allow for any possible reduction in strength during conversion. The advice of the manufacturers should be sought and their recommendations adhered to. Reference can be made to the brief notes on HAC in Chapter 1 and to the Bibliography at the end of this chapter.

The experience of the author suggests that for piles, the need to provide a very high slump, self-compacting concrete is likely to rule out the use of HAC. In the section which follows on the protection of concrete piles, the use of the new 'superplasticisers' with Portland cement concrete has been suggested. Some brief information on these materials is given in Chapter 1. An appreciable amount of work has been done on the possible long term effects of these new materials on the durability of Portland cement concrete and the results so far are very satisfactory, but the author has not found any published information on their use with HAC concrete.

The recommendations given in this chapter for the protection of concrete in foundations located in an industrial tip may appear to be exaggerated, but those who are responsible for such work should remember that the total cost of rectifying a failure is likely to be many times greater than the adoption of the protective measures when the foundations are originally constructed.

Mention is made of the possible use of an extra (sacrificial) thickness of concrete. This can be very effective, practical, and economic, but it is essential that this concrete should be of the highest quality and fully compacted.

6.2.4 PILED FOUNDATIONS

The author's experience is that piled foundations in an aggressive sub-soil generally present the greatest difficulty in protection because of the practical problem of how to actually provide the necessary protection to the concrete.

In the case of driven reinforced or prestressed piles, the solutions appear to lie in the following measures taken either singly or in some suitable combination which will depend largely on site conditions.

(a) a dense, impermeable concrete (which would be required in any case to withstand the driving stresses) with a minimum cover to reinforcement of 75 mm and characteristic strength 40 N/mm^2;
(b) an extra (sacrificial) thickness of concrete of not less than 75 mm;
(c) a protective layer of epoxide resin, fully bonded to the concrete, and formulated to have maximum resistance to abrasion, not less than 0·75 mm thick. In extreme cases, this layer may have to consist of an epoxide mortar.

With prestressed post-tensioned piles, the author considers it desirable to give special protection to the metal duct in which the cables run. If by any chance, the concrete becomes damaged or defective, and as a result aggressive chemicals penetrate to the duct, the duct casing will be rapidly attacked. While in theory, the duct should be completely filled with grout, this may not be the case in practice. Therefore the protection of the metal duct on the outside by a substantial coating of epoxide resin (say 0·50 mm thick) would be a valuable additional protective measure.

With bored piles the problems can be equally or even more difficult. The concrete has to have a very high slump, about 150 mm, in order to be self-compacting, and it must be sufficiently cohesive so as not to segregate when placed. In practical terms, this requires a cement content in the region of 400 kg/m^3. For durability, which is obviously of primary importance, the w/c ratio should not exceed 0·45. Such a mix would require a high grade plasticiser, but care must be taken with the dosage of plasticiser because if this is too high, there may be a permanent reduction in strength. All this means that trial mixes are essential. The use of one of the new 'superplasticisers' would be worth investigating. Brief information on these new materials is given in Chapter 1.

On the assumption that the above mix will ensure a dense, impermeable concrete in the pile (a result not necessarily achieved in practice, due to 'necking' and washing out of cement by ingress of ground water as the casing is withdrawn), the concrete may still require protection against chemical attack. This can be provided by one or more of the following:

(a) an extra (sacrificial) thickness of concrete, not less than 100 mm, *i.e.* the pile is 'oversized' by 200 mm on the diameter;
(b) a permanent casing of polypropylene pipe (provided a pipe of suitable diameter can be obtained); this casing must extend well below the lowest level of the tip material. An alternative is to provide a permanent steel casing, but the casing would have to be suitably protected on the outside by an abrasion and chemically resistant coating. The use of a casing may necessitate the neglect or substantial reduction of skin friction in the calculation of the bearing capacity of the piles.

6.2.5 Rafts, Strip Foundations and Column Bases

This type of foundation is generally easier to deal with than piles as the whole of the work is accessible. The object of the measures recommended is to isolate the structural concrete from the aggressive soil and ground water in those cases where the high quality of the structural concrete is not sufficient safeguard for long term durability. The protective measures can be one or more of the following:

(a) an extra (sacrificial) thickness of concrete on the base and sides of the foundation, not less than 100 mm;
(b) a protective layer of inert material under the bottom of the structural concrete, and carried up the sides to above the level of the aggressive soil or ground water. In this way complete tanking is formed. It is important to ensure that this tanking is free of pin holes and other similar defects which would admit aggressive water to the concrete. The materials used should be either insitu or preformed sheeting. The thickness of the tanking would depend on the materials used and the aggressiveness of the external conditions. Generally, sheeting should not be less than 1·0 mm thick and coatings which are bonded to the concrete, not less than 0·75 mm thick. All joints in sheets should be welded or preferably solvent-welded. If insitu coatings are used, the minimum number of coats

should be three and unless they are applied by spray, each coat should be applied at right angles to the preceding one in order to eliminate pin holes. Special care must be exercised at the junction of horizontal and vertical or sloping surfaces. The insitu materials which can be used for protection of concrete include mastic asphalt, heavy coats of bitumen reinforced with glassfibre mesh or polypropylene mesh, epoxide and polyurethane resins. The final selection would depend on chemical resistance, and ease of application under site conditions, ability of the coating to bond to the concrete, and physical resistance to damage by back-filling etc.

Mastic asphalt should comply with either BS 988 (BS 1097): Mastic Asphalt for Tanking and Damp-Proof Courses (Limestone Aggregate), or BS 1162 (BS 1418): Mastic Asphalt for Tanking and Damp-Proof Courses (Natural Rock Asphalt). On horizontal and sloping surfaces not exceeding 30° to the horizontal, the asphalt should be laid in three coats to a total thickness of 30 mm. On vertical surfaces and slopes exceeding 30°, it should be laid in three coats to a total thickness of 20 mm. On vertical surfaces, it should be bonded to the concrete as far as this is practical by thoroughly wire brushing the concrete prior to application of the asphalt. In addition, the mastic asphalt must be supported by a skin walling (which will also protect it from damage by back-filling). The walling is built up clear of the asphalt and the space left must be solidly grouted up with a cement–sand mortar, course by course as the work proceeds.

Bitumen reinforced with glass fibres or polypropylene mesh can form an effective barrier, but it does not bond well to damp concrete, and requires some physical protection against damage by back-filling. The physical protection can be exterior quality plywood. The problem from a long term durability point of view, in using asphalt or bitumen, is that if water does penetrate behind it, it can result in complete debonding from the base concrete, as the initial bond is rather poor.

Other insitu materials are epoxide resins and polyurethane resins. In some formulations these two resins are combined. These polymers can be formulated so as to bond well to damp and wet concrete and are resistant to physical damage. However, if the back-filling is granular material it would be advisable to provide temporary protection by means of thin plywood or at least hardboard. The resins should be applied in not less than three coats to a total thickness of 0·75 mm. On vertical surfaces and horizontal surfaces where there is no over-layer of concrete, the surface of the concrete must be carefully prepared by thorough wire brushing, light grit blasting or high velocity water jets (for information on the latter, see Chapter 5). The plastics sheeting includes 1000 gauge polyethylene plain, or with a bituminous-rubber adhesive on one side (proprietary name 'Bitu-Thene'), butyl rubber, 'Hypalon' (a synthetic rubber), polyisobutylene or PVC. The thickness should vary according to site conditions, but generally should not be less than 1·0 mm. Protection against subsequent operations such as back-filling, etc. is most important.

6.3 STRUCTURES HOLDING MILDLY AGGRESSIVE LIQUIDS

The structures considered in this section are those holding potable water derived from upland gathering grounds, contact tanks in water treatment works and tanks in sewage treatment works containing a mixture of domestic sewage and trade effluent. The recommendations given can also be applied to aqueducts and tunnels, and to trade effluent tanks in those cases where the effluent is only mildly aggressive to Portland cement concrete.

6.3.1 WATER RESERVOIRS

Water from upland gathering grounds usually contains organic acids and dissolved carbon dioxide, which may reduce the pH to 4·0; the usual range is about 4·5 to 6·5. These waters are also characterised by low temporary and permanent hardeners and low total dissolved solids (TDS). Some upland waters contain sulphurous and sulphuric acids which are formed from sulphur compounds in the sub-soil by bacterial action. In equal concentrations, these mineral acids are much more aggressive than the organic (humic) acids which are usually present. This is why it is important to obtain an adequate number of chemical analyses over an extended period of time. In this way, not only can changes in the pH be compared, but also the concentration of the various acids which are responsible for these variations. The maximum, minimum and average concentrations of aggressive compounds over at least one annual cycle should be tabulated. In many cases it will be found that there are sharp changes in the chemical composition of the water during and immediately following periods of heavy rainfall.

The attack usually takes the form of etching of the surface of the concrete, that is, the cement matrix is dissolved and the coarse aggregate (if it is inert to dilute acids), is exposed. Sometimes this etching can be very severe. Fig. 6.2 shows concrete attacked by an upland water.

It should be noted that these acid waters when they are flowing through pipelines, channels, etc. or are stored in reservoirs, are likely to be appreciably more aggressive to Portland cement concrete than when the same water is in the ground (ground water) surrounding foundations. The reason for this is that the acids in the flowing water are being constantly renewed and the corrosion products are washed away thus exposing fresh concrete to attack.

To combat these aggressive conditions concrete or mortar of the highest quality should be specified and used. This means for the concrete, a cement

FIG. 6.2. Concrete etched by acid and water from moorland (courtesy: Cement and Concrete Association).

content of not less than 360 kg/m^3 and a w/c ratio not exceeding 0·45; this low w/c ratio may necessitate the use of a plasticiser. In addition, it is recommended that a good quality limestone aggregate is used in preference to flint gravel or igneous rock. The only exception to the use of limestone is when the chemistry of the water is such that it would attack the limestone selectively, but such cases are very rare. The concrete must be fully compacted and properly cured. It has been found that free carbon dioxide (sometimes known as aggressive CO_2), is more aggressive to Portland cement than humic acids.

The mix for mortar should be 1 part Portland cement to 3 parts clean sharp sand, with only sufficient water to secure a workable mix. A plasticiser can be used if found necessary. Wherever possible the use of gunite (a pneumatically applied mortar) is recommended in preference to a hand applied rendering. The reason for this is that when properly executed, the gunite will be denser, stronger, and more impermeable than any hand applied mortar. The cement, sand and water are projected onto the concrete at high velocity and the w/c ratio seldom exceeds 0·35. The use of gunite in the repair and waterproofing of water retaining structures has been dealt with in some detail in Chapter 4.

After repairing the concrete with new concrete, or hand applied rendering, or gunite, a decision has to be taken whether the new work should itself be protected by an inert coating. In many cases, an extra thickness of concrete or mortar which will act as a sacrificial layer, will be adequate. The author wishes to emphasise again the importance of ensuring high quality, density and low permeability. If this is in fact provided, and not merely specified, then it is likely to be surprisingly durable, unless mineral acids are present in significant concentrations. The thickness of the sacrificial layer should not be less than 50 mm but 75 mm is better. It is however appreciated that by adding 75 mm thickness of concrete or gunite to the inside of an aqueduct or tunnel, the carrying capacity is correspondingly reduced, unless the original diameter is increased.

Where it is felt that the use of a sacrificial layer of Portland cement concrete or mortar of acceptable thickness will not give adequate protection in the long term, then the use of HAC instead of Portland should be investigated and discussed with the manufacturers. Reference can be made to the short discussion on the properties of this cement in Chapters 1 and 5.

If it is considered that the attack will be too severe to be resisted over a long period of time by either Portland cement or HAC, then the provision of a protective lining of inert material must be investigated. If the structure holds or carries potable water, then the lining must be non-toxic, non-tainting and must not support bacterial or fungoid growth. It is advantageous if a clearance certificate can be obtained from a large water authority for the use of the material in drinking water reservoirs.

Materials likely to be suitable are epoxide resins for insitu application, butyl rubber and polyisobutylene, in the form of preformed sheeting. When sheeting is used, the joints should be lapped and solvent or heat welded, and it is usual to bond the sheets to the base concrete. Bituminous emulsions (drinking water quality where appropriate), can be used for insitu application, but they do not bond well to damp or wet concrete, whereas epoxide resins can be formulated to give a strong bond even under conditions of extreme wetness.

Information on the application of these materials has been given in Chapter 4 to which reference should be made. It is important to mention that if preformed sheeting is used, special care should be exercised to ensure that it is not stopped off below top water level. Although the adhesives used are classified as not water-sensitive, if an exposed edge is submerged for a long period of time, some weakening of the adhesion is likely to occur. For this reason, when it is required to repair and protect only part of the surface, an insitu material is likely to prove more economic unless there are special reasons, such as maximum flexibility, which require the use of plastics sheeting.

6.3.2 Contact Tanks and Sewage Tanks Holding Trade Effluent

Structures which can be considered as falling within the category dealt with in this section are contact tanks in water treatment works and sewage tanks which contain a mixture of town sewage and trade wastes.

In contact tanks, chemicals are added to the water to assist in the treatment process. These chemicals include such compounds as aluminium sulphate (alum), ferrous sulphate, ammonium compounds and sulphuric acid. All these are aggressive to Portland cement concrete, the aggressiveness depending largely on the concentration.

When dealing with this type of structure it is essential to obtain detailed information on the operating procedure of the treatment process and the concentration of the various chemicals used, as well as the pH range. In public water supplies, the treatment operates at ambient temperature, but contact and neutralisation tanks in industrial plants may hold the solutions at elevated temperature. All these details must be clearly established before a satisfactory protective system can be decided upon.

While high quality, dense impermeable concrete made with sulphate-resisting Portland cement may be satisfactory for many tanks, there may well be others where inert protective linings may be required. For intermediate conditions, an extra (sacrificial) thickness of Portland cement concrete or gunite may provide adequate protection; as previously mentioned, consideration can usefully be given to HAC gunite.

The discharge from certain types of demineralisation plants can be extremely aggressive to Portland cement and HAC, and will be briefly considered in a later section.

Sewage tanks which contain a mixture of domestic sewage and trade effluent can be particularly difficult to deal with because of the variation in the chemical composition of the trade wastes. A great deal depends on the dilution factor. Trade wastes are often discharged at night when the dry weather flow in the sewers is at its lowest and sometimes these wastes are at a high temperature which increases their chemical activity.

It is of particular importance that the fullest possible information be obtained on the volume, period of discharge, temperature and chemical composition of the trade wastes. All relevant factors must be taken into account if a practical and economic solution is to be arrived at. The basic principles outlined in the earlier part of this chapter should be applied.

Problems of aggression to concrete in the sewerage system itself, particularly that arising from two stage bacterial action resulting in the formation of sulphuric acid, are dealt with in a later section.

6.3.3 Sludge Digestion Tanks

Tanks for holding digesting sludge at elevated temperature are dealt with here rather than in Chapter 4 because of certain special features in their operation.

Sludge digesters are now normally included in all sewage treatment works. They are usually circular on plan with domed roofs. The walls and floor are reinforced concrete (sometimes the walls are prestressed), and the roof is either heavily galvanised steel or reinforced concrete or reinforced gunite. Heating and mixing equipment is provided inside the tank. It is usual practice to maintain the temperature of the sludge in the range 25°–30°C, and optimum pH is 7·0–7·8. This means that the sludge in the tank should be either neutral or alkaline. The considerable volume of gas given off by the digesting sludge (about 23 litres per head of population) consists of about 70% methane and 25% carbon dioxide, the remaining 5% containing hydrogen sulphide and other gases.

It is important that the whole of the structure above average sludge level should be reasonably gas-tight, and the walls and floor should be liquid tight. While concrete is not considered a gas-tight material, high quality, dense impermeable concrete does perform satisfactorily in sludge digestion tanks as the gas pressure is very low. However should trouble develop through loss of gas, the joints should be first checked; if these are found in order and the concrete itself is at fault, then two coats of a suitable paint system on the inside of the roof and upper part of the walls should cure the trouble. The coating material selected must be durable under the operating conditions, which include cyclic thermal movements in the structure. With an inside temperature of 30°C and an outside temperature as low as −10°C, there will be an appreciable thermal gradient through the walls and roof. This temperature difference between inside and outside will be subject to fairly rapid change as the ambient air temperature fluctuates according to climatic conditions.

This book is not concerned with the structural design of these tanks except in so far as this has a bearing on how defects occur and finding a satisfactory durable method of repair.

Typical defects in sludge digesters are: (a) loss of gas and (b) chemical attack on the concrete. There may of course be other defects, such as cracks and defective joints, which are applicable to reinforced concrete structures in general, but these have been dealt with in other chapters of this book.

At this point the author would emphasise that strict safety measures must be taken when men enter sludge digesters for inspection etc. due to the possible presence of dangerous gases such as methane, carbon dioxide, hydrogen sulphide etc.

Loss (Leakage) of Gas

Areas and points of gas leakage can be traced by the use of soap solution applied to the outside. It is better to seal permeable areas of concrete and defective joints on the inside, but this will necessitate putting the digester out of commission for a period, and may not always be practical. In such a case, outside sealing must be adopted, and to do this, the gas pressure should be reduced as much as possible (it is not high in any event and seldom exceeds 6–8 in water gauge or 0·015–0·02 atm). Where the gas is seeping through permeable (honeycombed) areas of concrete, this can be either pressure grouted, as briefly described in Chapter 4, or sealed by surface application of grout or resin. In both cases, the first coat of the material used should be rapid setting, particularly for the surface application.

Cracks are more difficult to deal with because of the thermal movements taking place in the structure, which have been mentioned earlier in this section. Crack repair techniques have been discussed in some detail in Chapter 3, and the principles laid down are applicable here. The material used for injecting into the cracks or for sealing cracks which may have been cut out, may have to possess an adequate degree of flexibility, and must be durable when in contact with sludge gas.

For sealing joints, preformed neoprene gaskets are likely to be the most durable, but the joint sides must be true and smooth. To overcome this difficulty of irregularity in existing joint grooves, a channel shaped neoprene gasket has recently been put on the UK market and is shown in Fig. 4.20 (p. 165).

Chemical Attack on the Concrete

This is likely to show itself initially by some seepage through joints and/or cracks in the walls, usually accompanied by rust stains due to corrosion of the reinforcement. It is only after the emptying and thorough cleaning of the inside of the digester that the extent of the damage can be ascertained and methods of repair determined. The use of high velocity water jets for this work is recommended; for details of this method the reader is referred to Chapter 5.

Except in rare cases involving the serious malfunctioning of the digester, the attack is likely to show itself as shallow or deep etching of the surface of the concrete, with possibly increased deterioration at joints and cracks. For satisfactory digestion, the sludge in the tank should be maintained slightly above a pH of 7·0. However, the incoming raw sludge may be as low as 6·0.

The author's experience is that when attack does occur it is usually at or above average sludge level. All indications are that this is caused by sulphuric acid produced by two stage bacterial action which converts hydrogen sulphide in the sludge gas to H_2SO_4. Details of this process are given later in this chapter. Basically, the hydrogen sulphide is formed by anaerobic bacteria in the digesting sludge, and then the gas is converted to sulphuric acid by aerobic bacteria in the presence of moisture. Theoretically, there should be no oxygen in the space above the digesting sludge, but in practice there is sufficient oxygen to enable aerobic bacteria to flourish, at least from time to time.

This may cause some surprise to managers of sewage treatment works because the quantity of hydrogen sulphide (H_2S) in sludge gas is very low. However, there is little doubt that this conversion of H_2S to H_2SO_4 does sometimes occur, usually intermittently. The author has found quantities of a yellow powder on the inside of the roof of sludge digesters and on analysis this was found to be 70% elemental sulphur. Sulphur and H_2S are themselves not aggressive to concrete. Recommendations for protecting the concrete in these circumstances are given later in this chapter in the section dealing with repair and protection of sewer pipelines.

6.3.4 Precipitators for Flue Gases

Electrostatic precipitators are used in a number of industries where the flue gases contain large volumes of fine particles which for one reason or another have to be removed before the gases are released to the atmosphere.

Some flue gases contain sulphur and other compounds which are aggressive to concrete. Generally, no damage is done as long as the temperature of the gas is well above the dew point so that no liquid is present. As previously stated in this book, chemicals in the dry state seldom attack concrete. When damage does occur in these installations it is usually at the outlet where the temperature is lowest, and attack is likely to take place slowly over a period of several years.

The chemicals causing the attack must be identified before suitable remedial measures can be decided upon. If, as is often the case, the aggression is found to be a combination of acid, sulphate and ammonium attack, then HAC gunite, as is used in industrial chimney linings, is likely to be effective. Some general information on this cement is given in Chapter 1 and prospective users should seek advice from the manufacturers, Lafarge Aluminous Cement Company. The repair of buildings and water-retaining structures with gunite has been described in some detail in Chapters 3 and 4, and the principles laid down there are applicable here.

For the protection of undamaged Portland cement concrete, the use of a proprietary chemical and heat-resistant compound, such as that used in the flues at the Drax Power Station, and described later in this chapter, can be considered.

6.4 TUNNELS THROUGH INDUSTRIAL TIPS

Earlier in this chapter the problems of concrete foundations in industrial tips were discussed from the point of view of protecting the concrete against attack. Tunnels have to be lined, and except when they go through solid rock, the lining consists of either cast iron bolted segments, precast concrete segments (bolted or plain), or insitu concrete. All these materials require protection when the ground surrounding the tunnel is aggressive. As already mentioned, industrial tips may contain a wide range of aggressive chemicals and therefore it is the outside of the lining which needs to be protected. It is obvious that significant deterioration of a tunnel lining is a very serious matter indeed. Usually tunnel sewers are expected to have a life of not less than 80–100 years.

It is usual in tunnels to pressure grout behind the lining and special grouts can be formulated which will be resistant to many types of chemical compounds. However, it must be realised that pressure grouting cannot be relied upon to provide a complete external envelope, nor to achieve a 100% penetration of the ground surrounding the tunnel. While the area of contact between the outside of the lining and the aggressive ground/ground water may be very substantially reduced, there may still be areas which are not completely sealed off.

Measures which can be used in addition to pressure grouting, are:

(a) the use of HAC concrete instead of Portland cement, provided this is considered suitable after discussion with the manufacturers of the HAC;
(b) the provision of an extra (sacrificial) thickness of concrete;
(c) the provision of a protective layer of epoxide resin mortar on the outside and contact surfaces of the precast concrete segmental lining; this method of course only applies where precast concrete is used.

The measures outlined above are only relevant to new tunnels, so that the question naturally arises as to what can be done when an existing tunnel lining shows signs of deterioration due to attack from aggressive ground water. In such a case, pressure grouting would appear to be the only practical solution apart from renewal of the lining and replacement with a more durable material. Even in the latter case, pressure grouting should be used to provide additional protection. The first step would be a detailed subsurface investigation to try to define as accurately as possible the source of the aggressive chemicals and the boundary, both in plan and depth, of the contaminated ground. For example, there may be special site conditions which would indicate that the construction of a cut-off wall would substantially reduce the flow of aggressive ground water along the line of the tunnel. If this cut-off were combined with chemical injection into the ground between the cut-off and the line of the tunnel, a very real improvement may be effected. Some information on the waterproofing of tunnels is given in Chapter 4.

6.5 SEWER PIPELINES AND ANCILLARY STRUCTURES

The information, recommendations and conclusions given are just as relevant to asbestos cement pipes as they are to concrete pipes. The term 'ancillary structures' is intended to include manholes and sumps of pumping stations, but not tanks at sewage treatment works as these latter structures have been dealt with in a previous section (Structures holding mildly aggressive liquids).

Normal domestic sewage is not aggressive to Portland cement concrete. This is proved by the fact that out of the hundreds of thousands of miles of public sewers existing in the UK, the percentage which have suffered chemical attack to any significant degree is very small indeed.

However, when trouble does occur it is usually very costly to remedy. When concrete in sewerage systems suffers chemical attack it is due to one of three basic causes:

1. aggressive chemicals in the sewage (*e.g.* from trade wastes);
2. aggressive chemicals in the ground or ground water surrounding the sewerage structure;
3. the creation of aggressive chemicals within the sewerage system by what is termed 'two-stage bacterial action'.

Causes 1 and 2 above have already been discussed in previous sections of this chapter.

6.5.1 The Creation of Aggressive Conditions Within the Sewerage System by Two-Stage Bacterial Action

Under conditions of low velocity (*i.e.* a velocity less than is required for the sewer to be self-cleansing), often combined with retention of the sewage in sumps of pumping stations because the sewage has to be pumped several times, hydrogen sulphide (sulphuretted hydrogen) may be generated in the slimes and deposits in the invert of the sewers and sumps by anaerobic bacteria. The bacteria generate the gas by decomposing sulphur compounds in the sewage. At one time it was thought that the presence of trade effluents containing high concentrations of sulphur compounds was necessary for the initiation of the bacterial action. However, it is now known that the formation of hydrogen sulphide (H_2S) in this way can occur in domestic sewage. The presence of higher than normal concentration of sulphur compounds will provide additional food for the anaerobic bacteria and thus result in larger volumes of H_2S being formed.

Hydrogen sulphide on its own will not attack good quality concrete, but when the gas is formed and cannot escape quickly from the sewerage system, it may be converted to sulphuric acid by aerobic bacteria in the moisture saturated air above the water line in sewers and sumps of pumping stations. Sulphuric acid is very aggressive to both Portland cement and HAC concrete.

Until a few years ago it was thought that this two-stage bacterial action was confined to countries with a warmer climate than that of the UK, unless very special circumstances existed. An example of these 'special circumstances' was the sewers at Burton-on-Trent, which carried large quantities of brewery wastes. This particular trade effluent contained high concentrations of sulphur compounds. Since then however, cases of this type of attack have come to light in sewers in the UK which carry only domestic sewage, and in other countries with a similar, or colder climate such as Finland.

The attack develops above the average water level in the sewer pipe lines, sumps of pumping stations, and manholes. It should be made clear that the higher the temperature in the sewers the greater danger there is of attack. Normal fresh sewage is unlikely to generate hydrogen sulphide (which is the first stage in this form of attack), but if the sewage becomes septic, and this also applies to deposits in sewers and sumps, then the danger of H_2S generation is greatly increased. Septicity in sewage is a biological process caused by the activity of anaerobic bacteria as the result of which certain organic compounds are reduced from higher to lower conditions of oxidation, some of the solid organic substances are rendered soluble, and a quantity of gas is given off; this gas consists of carbon dioxide, methane, and hydrogen sulphide. Countries which have experienced severe aggression in their sewerage systems include Australia, South Africa, Southern USA, Egypt and Malaya.

It was C. D. Parker, in Australia, in about 1945, who was responsible for clearly defining the mechanism of this form of attack, *i.e.* presence of sulphur compounds (which are always present in sewage), suitable environmental conditions for the breeding of the anaerobic bacteria which reduce these compounds leading to the formation of H_2S, and the conversion of the H_2S to sulphuric acid (H_2SO_4), by the action of aerobic bacteria.

The principal factors involved in this process are:

(a) The temperature inside the structures (sumps, tanks, etc.) plays an important part. It is considered that the temperature favourable for this type of attack is generally within the range of 12–30 °C. The effect of temperature in the upper two-thirds of this range is very marked and H_2S generation rises steeply above 16 °C.

(b) Work in South Africa has shown that H_2S production is likely to reach a maximum at a pH of about 7·7 and falls sharply when the pH rises above 8·0. It should be noted that the pH of average town sewage is likely to be in the range of 6·8–7·8.

(c) In sewage treatment, the biochemical oxygen demand (BOD) is important. While the effect of the BOD of the liquid is less marked than the temperature, it has been found that the total sulphides formed are likely to be directly proportional to the BOD.

(d) The emission of H_2S into the air above the liquid is considerably facilitated by turbulence of the liquid.

(e) Ventilation of the space above the liquid in locations where H_2S is formed can be a decisive factor. Even when large quantities of H_2S are formed, provided there is good ventilation and the gas is dispersed rapidly to the open air, there will be little chance of the formation of sulphuric acid, and so attack on the concrete will be eliminated, or substantially reduced.

This type of attack on concrete and asbestos-cement in sewerage systems can be recognised by the following characteristics:

(1) It occurs above the water line.
(2) The hardened cement matrix in contact with the aggressive agent (sulphuric acid), is reduced to a soft putty-like material, having a pH at or below the neutral point of 7·0 instead of 10·5–11·5 in normal concrete.
(3) If this material is analysed it will be found to consist largely of calcium sulphate.
(4) Under certain circumstances, yellow deposits of a powdery material are formed on concrete and metal surfaces in contact with the hydrogen sulphide gas. On analysis this material will be found to consist of about 70 % elemental sulphur. It is apparently formed by a complex chemical-microbiological reaction.
(5) The effect of the attack, if it has been proceeding more or less continuously for an appreciable period, can be very serious. A pipe wall can be reduced in thickness by 75 % so that its strength is reduced almost to zero.

FIG. 6.3. Concrete attacked by sulphuric acid.

Figure 6.3 shows concrete attacked by sulphuric acid formed by two-stage bacterial action.

The question now arises as to what can be done to protect concrete from this type of attack. The author feels that prevention is always better than cure and therefore consideration will first be given to measures which will prevent conditions arising which are favourable to the formation of the anaerobic–hydrogen sulphide–aerobic–sulphuric acid cycle.

New Sewerage Systems

Measures should be taken during the design of the system to reduce turbulence to a minimum or to produce it under controlled conditions so that H_2S given off can be vented to the open air. Alternatively if this cannot be done, then special measures may have to be taken to protect the Portland cement concrete near the areas of turbulence. Experience in Australia and South Africa has shown that turbulence plays a very important role in producing aggressive conditions. This is logical because turbulence releases the H_2S, and it is the H_2S in the air space above the water level which is converted to H_2SO_4.

Every effort should be made to provide the best possible ventilation to remove the H_2S away from the humid warm environment inside the sewer. The writer has seen a case where the replacement of a sealed cover by a grating on an inlet chamber to a sludge tank resulted in the complete cessation of attack.

Special attention should be paid to obtaining self-cleansing velocities in all pipelines. If this is not practical, as may be the case in new systems where the initial flow may be very much less than the final design flow, provision should be made for regular flushing and cleaning to remove accumulation of slime, etc. from the invert. It is in these deposits that the anaerobic H_2S producing bacteria flourish. Another advantage of adequate flushing is that it removes any accumulation of acid which might have formed on or above the water lines. Unfortunately the beneficial results of these measures are not very long lasting and generally should be repeated at about fortnightly intervals.

In the case of a proposed scheme where attack in the early years is assumed to be likely, adequate protection can usually be given by coating the concrete with a specially formulated epoxide resin at least 0·6 mm thick. Such a thickness would require at least three coats. It is essential that there should be full bond between the coating and the concrete, and that there are no pin-holes. In many cases of failure of such coatings the primary cause has been loss of bond between the coating and the concrete due to inadequate preparation of the concrete surface. Contributory factors are incorrect formulation of the resin, inadequate film thickness and careless application resulting in pin-holes.

The need for proper preparation of the base concrete can raise serious practical problems when dealing with pipes which are too small to be entered by workmen. This preparatory work usually involves grit blasting, high velocity water jetting or thorough wire brushing.

Equipment has been developed in the UK for internally coating pipes in the field as well as individual pipes in the factory or on site, with epoxide and polyurethane resins. Figure 6.4 shows this equipment and Fig. 6.5 shows pipes being lined on site.

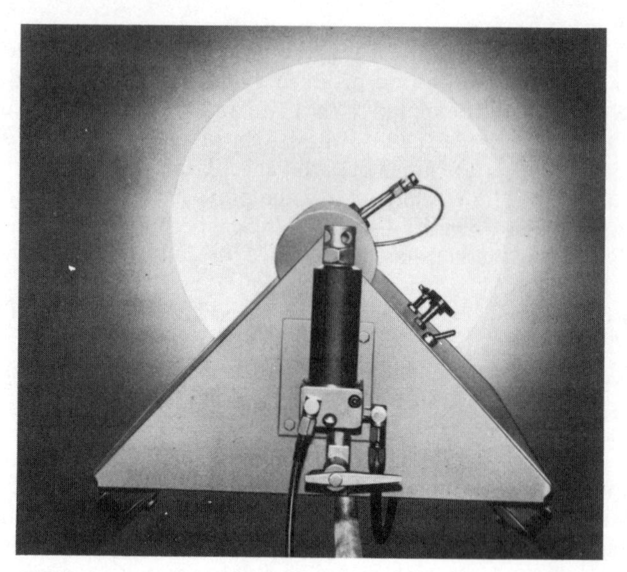

FIG. 6.4. Equipment for lining pipes with epoxide resin (courtesy: Colebrand Ltd).

As an alternative to epoxide resin lining, the patent method of impregnating the concrete with silicon tetrafluoride gas can be used providing the necessary equipment is available. This technique is known as 'Ocrating' and originated in Holland, and is used under licence in a number of countries. The success or failure of this method depends on several factors of which the principal are:

(a) the depth of penetration of the gas (SiF_4) into the surface of the concrete; a minimum depth of 5 mm is required;
(b) the concentration and quantity of sulphuric acid present.

The protection given to the concrete appears to arise from the formation of calcium fluoride by the reaction between the SiF_4 gas and the hydration products, principally calcium hydroxide, $Ca(OH)_2$, in the cement paste.

FIG. 6.5. Epoxide resin lining to concrete pipes (courtesy: Colebrand Ltd).

The silicon tetrafluoride gas is extremely active chemically; it is highly corrosive to most metals and is poisonous to human beings. These characteristics add to the difficulties of application. Experience shows that ocrated concrete possesses greater resistance to attack than unprotected Portland cement concrete. In South Africa it was found that the use of a limestone aggregate increased durability still further. The use of ocrated concrete may therefore bridge the time gap which exists between the construction of the sewerage system when flows are very small and when this type of acid attack is more likely to occur, and later when there is sufficient flow in the sewer to make conditions unsuitable for the generation of large quantities of hydrogen sulphide and its conversion to sulphuric acid.

Another solution is to use a new material known as 'F.70' which the author has seen applied successfully to the outside of cement silos to provide an impervious surface to the concrete after extensive repairs. The F.70 is described by the makers as a modified orthosilicate solution, and for use in sewers to withstand sulphuric acid attack, it would be reinforced with finely divided (powdered) stainless steel. This coating is inert to a wide range of chemicals including a number of dilute acids. While the makers would not at present be prepared to guarantee it against unspecified concentrations of sulphuric acid, it should give an appreciable measure of protection during the early life of the sewerage system.

Preformed linings of fibreglass impregnated with a polymer resin are available on the market. One of the disadvantages of this method of protection is that it is difficult to ensure completely gas-tight joints between the sections of preformed lining. If hydrogen sulphide seeps through the lining or joints and is converted to sulphuric acid, attack on the concrete pipes can proceed undetected until the pipes are badly damaged. Any form of lining which is not fully bonded to the base concrete is potentially dangerous for the same reason.

Research extending over many years in South Africa has shown that the use of high grade limestone aggregate results in an appreciably longer life than when the concrete is made with a flint gravel or igneous aggregate. It is important that the limestone aggregate should have a similar degree of permeability to that possessed by the cement paste. The reason for the use of limestone aggregate is that the attack from the acid will then be spread over both the cement paste and the aggregate instead of the acid being neutralised by the limited amount of cement paste alone. Therefore it may be advantageous in new schemes where it is thought that this form of attack may develop, to use a good quality limestone aggregate in the concrete.

Apart from the type of aggregate used, the first line of defence in any concrete structure against aggressive attack irrespective of what form it takes, is to provide a dense impermeable concrete. This can be achieved by an adequate cement content, not less than 360 kg/m^3 and a low w/c ratio (maximum 0·45); in addition to this, the concrete, when placed, must be fully compacted in order to obtain maximum density and be carefully cured.

Concrete sewer pipes made in accordance with BS 556 would come within this category of high grade concrete as the cement content is unlikely to be less than 400 kg/m^3 and irrespective of the method of manufacture, the final w/c ratio is likely to be not more than 0·40. Some manufacturers claim a cement content of about 450 kg/m^3 and a w/c of 0·33. Concrete of this quality is required in order to meet the stringent testing conditions imposed by the British Standard. Asbestos cement pipes, complying with BS 486 Pressure Pipes and BS 3656 Sewer Pipes are also entirely comparable with high grade concrete pipes.

Existing Sewerage Systems

The problems which arise with this type of attack in existing sewerage systems are in many respects more difficult to deal with than in new systems. Probably the most difficult one is to ascertain with reasonable accuracy the extent of the damage inside relatively small diameter pipelines (150–900 mm), which cannot be entered by men.

When trouble is suspected, the first step is to inspect the sumps of pumping stations, and manholes at the discharge end of pumping mains as it is here that attack is most likely to occur. When deterioration is found in these manholes, the investigation should be extended into the top lengths of the rising mains and the upper sections of the gravity sewers taking the discharge. When the diameter of the pipelines is less than about 900 mm, the inspection is best carried out by closed circuit television. It is advisable to clean and flush out well each length of sewer before the TV survey is commenced. Unfortunately, even the best photographs do not show the depth of attack on the concrete. They will show that corrosion has taken place, but it will not be possible to determine from the TV survey alone whether the pipe wall has been reduced in thickness to an extent likely to effect the load bearing capacity of the pipes. To resolve this vital question it will be necessary to excavate and remove some of the pipes for detailed examination.

The author believes that it should be possible for interested firms to develop equipment to record significant changes in pipe diameter insitu in the field.

All practical steps should be taken to reduce turbulence and to provide ventilation at places where turbulence occurs. Improved ventilation to all sections of the sewerage system where hydrogen sulphide is known to be forming will give beneficial results.

The repairs to damaged and deteriorated concrete should be carried out on the following lines:

(a) All deteriorated and softened concrete should be carefully removed, followed by thorough washing with clean water.

(b) It is advisable for the new concrete or mortar to be made with a styrene butadiene based latex emulsion, and placed in accordance with the principles set out in this book.

(c) After the new concrete or mortar has cured and hardened, at least three coats of a high grade epoxide resin should be applied to a total thickness of not less than 0·60 mm, or three coats of the modified orthosilicate solution reinforced with stainless steel powder referred to earlier in this section. It is likely that the epoxide resin will provide protection over a longer period than the orthosilicate solution. The selection of the protective coating should take into account the probable reduction in the aggressive conditions by the means suggested in this section.

Arrangements should be made for the regular cleaning of the pipes, rising mains and sumps to reduce accumulation of slime, and to wash away acid which has formed on the concrete.

Treatment of the sewage itself will often yield very beneficial results by reducing the formation of hydrogen sulphide and inhibiting the activity of anaerobic and aerobic bacteria. Work in South Africa showed that adequate dosing with lime will raise the pH to between 8·0 and 9·0. It has been stated previously that the formation of hydrogen sulphide is substantially reduced when the pH of the sewage is above 8·0. The South African experience indicated that dosing with lime was cheaper and more effective than chloride of lime which released chlorine. However, this result is not necessarily applicable to all cases. Other reports indicate that injection of chlorine at the pump outlet is effective. Some authorities prefer to dose the sumps with lime and follow this with chlorine injection into the pumping main. The injection of oxygen into the pumping main has also been successful.

A new method of treatment is the addition of hydrogen peroxide (H_2O_2) into the sewage at suitable points. The hydrogen peroxide reacts with the hydrogen sulphide to form water and sulphur, and the latter settles out as a deposit and is not aggressive to Portland cement concrete. A report on the development of this method was given in the journal Water and Sewerage Works (USA), in August 1973. Subsequent work elsewhere confirms the effectiveness of this method in reducing the formation of hydrogen sulphide and consequently the final formation of sulphuric acid. However, practical trials suggest that it is much more expensive than the other methods mentioned above.

As temperature plays an important role in the formation of hydrogen sulphide (H_2S generation rises steeply above 16 °C), every practical effort should be made to prevent the admission of high temperature effluents to the sewerage system.

The final recommendation is that careful and regular inspections, say twice a year, should be made to monitor the effect of the remedial measures which have been taken. Where considerable operating expense is involved, it may be found that less costly techniques give reasonably satisfactory results.

6.6 EFFLUENT FROM DEMINERALISATION PLANTS

Chemicals used for the regeneration of ion exchange demineralisation plants can be very aggressive to concrete. These chemicals include concentrated sulphuric acid and caustic soda, and the resulting effluent can have a temperature in the range 10–60 °C. These strong solutions are usually discharged to a neutralisation sump separately, and if by mischance, the concentrated acid solution is discharged before neutralisation, severe damage can be caused to the drainage system. When the chemicals mentioned above are used, even the neutralised effluent will be very aggressive to Portland cement (*e.g.* $H_2SO_4 + 2NaOH = Na_2SO_4 + 2H_2O$). Depending on the sulphate ion concentration, the use of sulphate resisting Portland cement may not provide adequate protection, and inert coatings to the pipes and insides of manholes, etc. will be required. The neutralisation sump will obviously require carefully thought out protective measures, particularly as sand may be present, and will act as an abrasive.

The solution will lie in the use of chemically resistant clay bricks and tiles bedded and jointed in a chemically resistant mortar, or a specially formulated epoxide/polyurethane resin system. The protection of the drainage system (pipes and manholes) is also important. This protection is likely to involve the interior lining of the pipes, and coating the inside of the manholes to about 1·0 m above invert level with a material which will withstand attack even from the unneutralised effluent. The capital cost will be high unless the length of pipeline is short, but will be far less than having to renew the system after it has been in operation some years.

The following additional precautions are also recommended:

(1) Equipment should be installed in the neutralisation tank to ensure proper mixing. This can consist of either mechanical agitators or a grid of perforated pipes on the floor through which compressed air is discharged.

(2) A device should be installed so that the effluent can only be discharged after it has been neutralised, *i.e.* when the pH is above 7·0.

(3) A detailed programme of inspection of the system below the outlet from the neutralisation tank, as well as the tank, should be prepared and acted on.

6.7 SEA WATER DISTILLATION PLANTS—HOT BRINES AND DISTILLATE

The advent of the large multi-stage flash distillation plants in the hot arid regions of the world has produced many problems of corrosion of metals and concrete. The temperature range of the final distillate and the brines is a wide one, and figures of 40–140 °C (105–290 °F) have been given in technical papers on this subject. Of particular interest is an investigation by Graham and Backstrom described in a paper published by the American Concrete Institute (see Bibliography). Unfortunately no information was given in the paper on the chemical composition of the brines used in the tests; the total dissolved solids (TDS) varied from 73 000–175 000 ppm, and the temperature from 38–143 °C. The temperature of the brine had a greater effect on its corrosiveness than the TDS content. For full information on the tests the reader is referred to the paper itself.

No detailed recommendations were given on suitable protective measures for the concrete, apart from the mention of sacrificial layers of concrete. The tests showed that high quality concrete was suitable without a sacrificial layer for brines in the temperature range 40–90 °C. For higher temperatures a sacrificial layer was required.

In practical terms the solution will depend on the details of the plant and method of disposal of the brines, and their dilution and cooling. A sacrificial layer to cope with permanently corrosive conditions presumes that the layer may have to be renewed periodically, and therefore provision for closing down that particular section of the installation must be included in the design and allowed for in the operation of the plant.

As in all cases of aggressive conditions, the first line of defence is high quality, dense and impermeable concrete. The mix should be as recommended for marine structures (see Chapter 5), but sulphate-resisting Portland cement should be used. It had been found that the hot distillate, which had a total dissolved solids content close to zero, was very aggressive to the cement matrix as it rapidly leached out the hydration products, resulting in serious loss of strength, rapid increase in permeability, and finally in disintegration of the concrete.

The distillate from these plants should be treated with lime so as to bring the pH to above the neutral point of 7·0, say to 7·5, and to provide a TDS not lower than about 150 ppm. Steps should be taken to control the free (aggressive) carbon dioxide if this is present in the distillate. It is also advisable to reduce the temperature of the distillate to about 35–40 °C before it comes into contact with concrete or asbestos-cement, if this is at all practicable.

6.8 BLOW-DOWN PITS

The mixture of steam and nearly boiling water from boilers contains a wide range of chemicals some of which are very aggressive to Portland cement concrete.

The pit must be constructed so as to resist physically the wide temperature range (probably 0 °–100 °C) in which it will have to operate. The material which is in contact with the blow-down must have adequate chemical resistance as well. The best long term solution is probably to construct the pit in reinforced concrete and line it with chemically resistant bricks bedded in a chemically resistant mortar. A high quality engineering brick such as Staffordshire Blues or Accringtons in HAC mortar may also give a long life in most cases. Acid-resisting bricks and tiles should comply with BS 3679: Acid Resisting Bricks and Tiles.

An alternative, which would be appreciably cheaper but have a much shorter life, would be to apply two coats of HAC rendering to the inside of the reinforced concrete pit. The concrete should be grit blasted, scabbled or otherwise prepared so as to expose the coarse aggregate and thus form a good key for the rendering. All grit and dust must be removed before the first coat of rendering is applied. The first coat should be about 10–12 mm thick, with mix proportions of 1 part HAC to 3 parts sharp clean sand. This should be combed or scratched to form a key for the second coat, and protected against too rapid drying out. The second coat should not exceed 10 mm in thickness and the mix should be slightly weaker than the first coat. This coat must also be protected against rapid drying out. HAC should only be used after discussion with the manufacturers; some general information on this cement is given in Chapter 1.

6.9 PROTECTIVE LININGS TO INDUSTRIAL CHIMNEYS

Chimneys of all types require to be provided with a lining. Domestic chimneys will not be dealt with in this section, as it is intended only to provide basic information on linings for reinforced concrete industrial chimneys.

A concrete chimney is normally lined for its full height and the reasons for lining may be summarised as follows:

(a) to reduce the thermal gradient across the structural chimney shell, *i.e.* to ensure that the stresses in the concrete and reinforcement are within certain predetermined limits;
(b) to protect the concrete from the chemical action of corrosive combustion products which are principally mineral acids and sulphates;
(c) to protect the concrete from abrasion caused by small particles in the flue gases.

Linings are of various types, principally as follows:

Independent linings
Linings which are supported on bearings, normally corbelled out from the structural chimney
Bonded linings, such as reinforced gunite, or other materials, usually polymers, specially formulated for the purpose

A reasonable amount of information is available on independent brick linings and linings carried on supports which are corbelled out from the structural chimney shell. Therefore the information that follows relates to two basic types of bonded lining, namely reinforced gunite and special newly developed polymer materials.

It is now common practice to construct tall industrial chimneys within a wind shield when the chimney is a multi-flue structure. In these cases the flues themselves are lined but not the wind shield. Generally, the top third of a flue is the part which is subjected to the most severe chemical aggression due to the higher frequency of condensation occurring in this section.

Chimneys which are used intermittently are more vulnerable to attack than those used continuously.

The main aggressive compounds in flue gases are acids (principally sulphuric) and sulphates. The aggressive conditions would be greatly reduced if low sulphur content fuels could be used. Flue gases from the burning of higher sulphur content fuels are often passed through special equipment to reduce the concentration of sulphur compounds, but unfortunately, this in turn lowers the temperature of the gases and increases their moisture content.

6.9.1 Reinforced Gunite Linings

The gunite which is used for lining industrial chimneys is usually made with HAC and a refractory aggregate such as crushed firebrick. The gunite attains a high strength within 24 h and has resistance to dilute acids, sulphates etc., which is characteristic of mortar made with this type of cement. For operating temperatures above about 150 °C, it is important that preheating be carefully carried out and that the temperature rise does not exceed 50 °C per hour.

It is important to select the correct aggregate for the operating conditions inside the chimney. For example, silica sand is only suitable for temperatures up to about 220 °C because of its high coefficient of thermal expansion. Crushed firebrick has an upper temperature limit of about 1300 °C and lightweight aggregates of expanded clay and shale and vermiculite can be used up to about 950 °C. The advantage of crushed firebrick and particularly the expanded clays, is their low bulk density and consequently their high thermal insulation value.

The basic principles of guniting with HAC are similar to those where Portland cement is used. The mix proportions are usually 1:3 or 1:4 by volume but in place it is likely to be appreciably richer due to loss of aggregate in rebound. Admixtures should not be used with HAC. The w/c ratio should be in the range 0·33–0·35. The surface of the base concrete should be lightly grit blasted, or hand scabbled so as to remove all weak laitance. The use of high velocity water jets for this preparatory work has a lot to recommend it; further information on this method is given in Chapter 5.

It is usual to provide a minimum thickness of 50–75 mm of HAC gunite. If a light steel fabric is used, as is usually the case, it should be securely pinned to the prepared surface of the concrete, and there should be a minimum cover of 20 mm of gunite. However, it is not advisable to use reinforcement in the gunite if the temperature of the steel is likely to exceed 400 °C.

Due to the rapid evolution of heat of hydration of HAC, it is most important that it should be kept wet, or at least moist for 24 h after placing. The wet curing should start as soon as the gunite has stiffened sufficiently not to be damaged by a fine spray of water. Before using high alumina cement it is advisable to consult the manufacturers; some general information on this cement is given in Chapter 1. Figure 6.6 shows a reinforced HAC gunite lining being placed inside an industrial chimney.

Reinforced gunite made with ordinary Portland cement is used to strengthen existing chimneys. The outside of the chimney is first cleaned and prepared in the usual manner and then the reinforcement is securely pinned in position. The gunite is then placed; Figure 6.7 shows a chimney which has been strengthened in this way.

FIG. 6.6. Lining chimney with reinforced gunite (courtesy: Cement Gun Co. Ltd).

FIG. 6.7. Strengthening and protecting steel chimney with reinforced gunite (courtesy: Cement Gun Co. Ltd).

6.9.2 PROPRIETARY BONDED LININGS

The author is indebted to Colebrand Ltd for the information which follows which deals briefly with the lining of Flues Nos. 1 and 2 at the Drax Power Station, Yorkshire. The height of the chimney is 260 m and the area coated in one flue was 7500 m^2. The coating used is described as a fluoro-elastomer and was applied by airless spray to a total thickness of about 1·5 mm. The maximum expected gas temperature was about 204 °C and the maximum gas velocity approximately 17 m/sec. This thickness was built up by one priming coat and 24 subsequent coats of specially formulated resin.

The specification required the preparation of the concrete surface of the flue by grit blasting; this was followed by careful priming and filling in of all cavities with an epoxide putty prior to the application of the priming coat of elastomer. Continuous quality control of both materials and workmanship was carried out during the lining process. Figure 6.8 shows work proceeding inside one of the two reinforced concrete flues; Figure 6.9 is a view of the outside of the reinforced concrete wind shield.

FIG. 6.8. Lining R.C. flue at Drax Power Station with fluoro-elastomer coating (courtesy: Colebrand Ltd).

6.10 LINING CHEMICAL TANKS

Reinforced concrete is the natural material for the construction of liquid retaining structures, but the liquids contained in most chemical tanks are aggressive to both Portland cement and HAC concrete. Therefore the structural shell of the tank has to be provided with a protective lining.

The basic principles involved in the selection of protective linings have been set out in the introduction to this chapter. The introduction also gave

FIG. 6.9. View of wind shield to R.C. flues being lined with fluoro-elastomer coating (courtesy: Colebrand Ltd).

information on the probable effect on Portland cement concrete of a number of chemical compounds in general use. Some of these compounds are only mildly aggressive, while others attack the concrete vigorously and quickly cause serious damage. The provision of a durable and impermeable lining is therefore essential. The selection of the most suitable material for a given set of circumstances requires specialist advice involving the combined experience of a civil/structural engineer and an industrial chemist.

The first person to be involved in the design is usually the engineer. Although the protective lining is applied after the completion of the structural shell, the type of lining and method of application should be decided very early in the design process. It is obviously important for the engineer to understand the basic principles involved and to know where to get expert advice on the lining.

The materials which are in more general use for lining concrete tanks are chemically resistant bricks and tiles, epoxide, polyurethane, and other organic polymers, and plasticised PVC sheeting. General information on the lining of chemical tanks with the materials mentioned above, as well as glass, rubber, stainless steel and nickel, is given in British Standard Code of Practice, CP 3003: Parts 1 to 10.

Recently, epoxide resins which are slightly flexible have come on to the market, together with flexible polyurethane resins. A fully bonded insitu coating has, in some circumstances, advantages over a preformed sheeting.

An example of a proprietary material which has excellent heat and chemical resistance, is flexible, and bonds well to concrete, is the fluoro-elastomer used to line the two flues at the Drax Power Station and described briefly in the previous section.

As far as the author is aware, there is no record of the total area of concrete tanks protected by the various materials mentioned above. Information obtained from contractors who specialise in these linings suggests that chemically resistant clay bricks and tiles, bedded and jointed in a suitable chemically resistant mortar, are the most widely used lining at the present time.

The result of a failure of a protective lining can be serious, and for this reason it is considered good practice in some cases, to provide a continuous membrane between the structural shell and the inside lining. This membrane acts as a 'reserve' protective layer and must therefore be to a reasonable degree resistant to attack by the contents of the tank. The inner, brick lining resists chemical attack and protects the membrane from physical abrasion and the direct effect of high temperature. The lining of chemical tanks is highly specialised as the range of chemicals used in industry is very wide, and the liquid contents of the tanks are often maintained at high temperatures. The membranes used include such materials as lead, rubber, asphalt, plastics sheet material, glass reinforced plastics, epoxide and polyurethane resins.

The corrosion-resistant mortars in general use include HAC/sand mortars, silicate mortars based on sodium or potassium silicate solutions, inert fillers and hardening agents, resin bonded mortars and rubber latex mortars based on natural or artificial rubber, inert fillers and silica sand. The resin mortars are generally based on epoxide resins, polyester resins, phenolic resins, furane resin.

Detailed information is given in British Standard Code of Practice CP 3003: Lining of Vessels and Equipment for Chemical Purposes, Parts 1 to 10, and reference can be made to the Bibliography at the end of this chapter.

The suppliers of the various materials used for the lining and the membrane will provide information in the form of printed Data Sheets on the chemicals which can be safely used in contact with the particular materials they offer. The information given is usually related to the concentration and the temperature range. Cases can arise where the supplier has only limited information on the behaviour of the material under the service conditions specified. In this event, either some risk must be taken, or if time permits, it will be necessary to work out a series of accelerated tests. These accelerated tests are very difficult to devise if they are to give reliable results. The preparation of the test programme and the interpretation of results presents many pitfalls. The effect of shortening the time, which is the basis for the accelerated tests, can give surprising results, sometimes too optimistic and sometimes the reverse.

FIG. 6.10. Concrete tank for industrial waste products lined with chemically resistant tiles (courtesy: Prodorite Ltd).

Figures 6.10 and 6.11 show concrete tanks lined with chemically resistant tiles and bricks.

Where tanks hold particularly aggressive liquids the author considers that whenever practicable, provision should be made for inspection of the outside of the walls and the underside of the floor during operation. In this way, should leakage develop (and this possibility can never be completely ruled out), remedial measures can be put in hand at an early stage, before serious damage to the structure, or contamination of the sub-soil has occurred.

FIG. 6.11. Acid waste neutralising tank lined with chemically resistant bricks (courtesy: Prodorite Ltd).

6.10.1 Epoxide Resin Linings

Epoxide resins are used for protecting concrete against chemical attack. In addition, these linings are often used to prevent contamination of the liquid held in the tank by chemical compounds leached out of the concrete. Because of their very smooth impervious surfaces the linings are also useful in enabling the cleaning and desludging of tanks to be carried out quickly and efficiently.

With concrete tanks the epoxide resin used is normally a two-pack material and is applied insitu by brush roller or airless spray. It is essential that the concrete should be properly prepared to receive the resin as the majority of failures of epoxide resins are mainly due to failures of bond caused by inadequate or improper preparation of the concrete. The basic requirement is that any weak layer of laitance, and all dirt and other contamination must be removed.

The resins are usually pigmented and when used in chemical tanks it is important that the pigment itself should be resistant to the chemical compounds with which it is in contact.

A basic requirement is that the coating must be sufficiently thick to resist any abrasion and also to ensure that it is completely free of pin-holes. Depending on the use to which the tank is put a minimum thickness of 0·5 mm is usually required, and in some cases this may have to be increased to 1·0 mm or more. Coatings of this thickness are built up in a number of successive layers, each layer being allowed to cure and the solvent to evaporate from it before the next coat is applied.

Epoxide resins are resistant to a wide range of chemicals and the information given below is of a general character and each case should be investigated as a separate problem before it is finally decided to use a particular formulation.

Epoxide resins are resistant to most acids except those of a strongly oxidising nature. They possess excellent resistance to alkaline solutions. They possess only limited resistance to chlorine and hypochlorites and may be strongly attacked by liquid chlorine and strong solutions of hypochlorite. They are resistant to a wide range of solvents but should not be used if the tank holds chlorinated hydrocarbons or phenols.

The temperature effect on epoxide resins varies with the formulation but as a general statement naturally curing resins can be used at temperatures up to 90 °C. However, the coefficient of thermal expansion of epoxide resins is appreciably higher than that of concrete and therefore they are not suitable where there is likely to be a wide temperature range. Epoxide resins will generally not cure in temperatures below 5 °C, and this can cause delay in their application during cold weather.

Where high build coatings are required, *i.e.* where it is required to obtain the necessary thickness with the minimum number of coats, the use of coal tar epoxies can be very useful provided these are suitable for the particular duty involved.

6.10.2 Plasticised PVC Sheet Linings

The physical and chemical properties of plasticised PVC depend to a large extent on the proportion and type of plasticisers used; plasticisers are incorporated in the material to produce flexibility and softness as well as tensile strength and resilience. The percentage as well as the type of plasticiser can also have a profound effect on the chemical resistance of the PVC sheet. Some chemicals tend to leach out or combine with a particular plasticiser, and this can result in serious deterioration of the sheet material.

The maximum safe operating temperature in tanks lined with PVC sheeting is about 60 °C. Provided the constituents of the sheeting have been correctly formulated the material is resistant to a wide range of dilute acids and alkalis at ambient temperature. It is important that the 'design' temperature should not be exceeded. Therefore when considering the use of PVC sheeting it is essential to investigate the chemical composition of the liquid with which it will come in contact as well as the maximum operating temperature. Complete information must be passed to the manufacturers of the sheeting.

While in the case of water-retaining structures, PVC linings are often used as a 'loose bag', in chemical tanks it is normal practice to bond the sheeting to the concrete shell. It is important that all corners and arrisses should be rounded and that the interior surface of the concrete to which the sheeting will be bonded is smooth and free from rough areas and sharp projections. Weak laitance on the surface of the concrete should be removed by careful wire brushing. The concrete should be allowed to dry out as much as possible before the lining is applied. Any attempt to bond the sheeting to damp or wet concrete is likely to meet with failure.

Where the tank is situated below ground level and may be subjected to external water pressure, it is essential that the walls and floor should be completely watertight before the sheeting is applied. Unless this condition is complied with the external water pressure may force the lining out of position when the tank is empty. In such an eventuality it would be most unlikely that the lining could be restored to its original position.

6.11 STRUCTURES HOLDING INDUSTRIAL COOLING WATER

Many industries, including power stations, have arrangements for circulating cooling water, and for topping this up to make good losses by evaporation. Large quantities of water are involved and sometimes little attention is given in the design stage of the project to the possible effect of the water on the concrete containing vessel, pipelines etc.

The water used is often not of drinking water quality as it is taken from a river or canal and little or no treatment given to it. Due to evaporation within the system, there is a build-up of salts in solution, mostly sulphates and chlorides. The temperature of the water is well above ambient, and sometimes the pH falls appreciably below the neutral point of 7·0, which means that the water under this condition is acidic.

It has been stated previously in this book that all Portland cement concrete is potentially vulnerable to attack by acids; the degree of attack depends on many factors details of which are given earlier in this chapter. Under the circumstances described above cooling water can be very aggressive to Portland cement concrete. To prevent this happening, the following precautions should be taken, either individually or in some combination:

(a) The water should be treated so that the concentration of salts in solution does not reach dangerous levels.
(b) The concrete should be of the highest quality: minimum cement content 400 kg/m^3, and maximum w/c ratio 0·45, thoroughly compacted and properly cured.
(c) An adequate thickness of 'sacrificial' concrete or gunite should be provided.

(d) An inert coating of adequate thickness suitable for the operating conditions should be provided.
(e) HAC concrete could be used instead of Portland cement, as a monolithic topping or in a protective layer of gunite. The advice of the manufacturers should be obtained before deciding to use this cement.

6.12 THE PROTECTION OF INDUSTRIAL FLOORS

For a detailed consideration of the protection and repair of industrial concrete floors, the reader is referred to the book 'Floors—Construction and Finishes', details of which are given in the Bibliography at the end of this chapter. However, the author feels that a brief review of the main problems and some advice on how to deal with them may be useful.

The problem of the protection of concrete floors against chemical attack may be divided into two parts, *i.e.* where there is spillage of aggressive substances onto the floor during the industrial process and where the floor is used for the storage of materials which are themselves aggressive to concrete and steel reinforcement.

Spillage

In the case of spillage, it is usually sufficient if the floor in the spillage areas only is given protection. The protective measures can include a sacrificial topping of Portland or HAC concrete (whichever is more resistant to attack), or the use of high quality hydraulically pressed concrete slabs or tiles, which can also be considered as 'sacrificial'. In many cases it is impossible to predict with any degree of accuracy how severe the attack will be. It is then a great advantage if the floor surfacing is such that it can be readily repaired in small areas; if the damage is confined to a topping or similar, the structural slab is not affected. Precast slabs and tiles can be quickly and easily replaced so that the shut-down time is reduced to a minimum. Repairs to an insitu concrete topping can also be carried out over a week-end by the use of HAC or ultra rapid hardening Portland cement concrete or mortar. When the HAC mix is correctly proportioned and the concrete or mortar properly cured, almost maximum strength will be reached in 24 h.

HAC is rather more resistant to dilute acids and sulphates in solution than Portland cements. It should be noted that tiles and slabs must be carefully laid and even then it is sometimes found that the joints deteriorate quicker than the units themselves. Figure 6.12 shows acid-proof flooring.

Storage Areas

When aggressive, or potentially aggressive materials are stored on the floor

FIG. 6.12. Concrete floor protected against acids by chemically resistant tiles (courtesy: Prodorite Ltd).

in direct contact with the concrete, attack often occurs, much to the surprise of the user. He cannot understand why what he considers are dry fertilisers, bulk sugar, or fruit and vegetables, will attack good quality concrete. In theory, if both the material stored and the concrete are maintained in a completely dry condition, attack will not take place. In practice, these ideal conditions are never realised, as moisture in some form is always present. The crystals of many substances are deliquescent, and the layer next to the floor becomes a wet sticky mess which in fact hides the aggression which is taking place on the concrete, until serious damage has been done.

Sugar, and all ammonium compounds in the liquid phase will attack concrete. Common salt is not aggressive to concrete, but is very aggressive to ferrous metals; therefore any cracks or defects in the concrete surface may allow the salt solution to reach the reinforcement, with serious consequences. The juice from all kinds of fruit and vegetables is acid and will attack the cement paste in the concrete.

As in the case of spillage it is often impossible to predict how much deterioration will occur in a given period of time. The choice therefore lies between doing nothing and risking serious attack and consequent dislocation of the business while repairs are carried out, or the provision of a 'temporary' sacrificial surfacing of insitu or precast concrete, or a more expensive topping or coating which is specially designed (formulated) to resist the particular chemicals present.

BIBLIOGRAPHY

AMERICAN CONCRETE INSTITUTE, *Manual of Concrete Practice*. Part 3—Products and Processes, A.C.I., Detroit, 1972, pp. 515–13 to 515–73

BUILDING RESEARCH ESTABLISHMENT, *Concrete in sulphate bearing soils and ground waters*. Digest No. 174, HMSO, London, 1975, p. 4

PERKINS, P. H., *The Use of Portland Cement Concrete in Sulphate Bearing Ground and Ground Water of Low pH*. Cement and Concrete Association, London, ADS/30, Apr. 1976, p. 16

FORRESTER, J. A., Concrete corrosion induced by sulphate bacteria in a sewer. *The Surveyor*, **188** (3571), (1959), 31 October, pp. 881–884

MINISTRY OF TECHNOLOGY, WATER POLLUTION RESEARCH LABORATORY, Formation of sulphates in sewers, *Notes on Water Pollution*, March 1966, Note 32, HMSO, p. 4

SOUTH AFRICAN COUNCIL FOR SCIENTIFIC AND INDUSTRIAL RESEARCH—BUILDING RESEARCH ADVISORY COMMITTEE, Corrosion of concrete sewers. Series DN 12, Report No. 163. Published by the authors, Pretoria 1959, p. 236

BARNARD, J. L., Corrosion of sewers. South African Council for Scientific and Industrial Research. National Building Research Institute. Bulletin 45, Pretoria, 1967, p. 16

PARKER, C. D., A survey of the extent of aggressive H_2S conditions and sewer corrosion in a number of representative sewerage systems in Victoria, Australia. Water Science Laboratories, Carlton, Victoria, September 1968, p. 41

PARKER, C. D., Comparison of the chemical and microbiological durability of asbestos cement and concrete sewer pipes under a variety of aggressive conditions. Water Science Laboratories, Carlton, Victoria, April 1969, p. 43

THISTLETHWAITE, D. K. B., *Control of Sulphides in Sewerage Systems*. Butterworths, Sydney, Australia, 1972, p. 273

US ENVIRONMENTAL PROTECTION AGENCY, *Process Design Manual for Sulphide Control in Sanitary Sewerage Systems*, US Environmental Protection Agency. 1974

ROBSON, T. D., '*Shooting*' *Stack and Flue Linings with Aluminous Cement Mortars*. Lafarge Aluminous Cement Co., Ltd., London (Reprint), p. 4

GRAHAM, J. R. AND BACKSTROM, J. E., Influence of hot saline and distilled waters on concrete. American Concrete Institute 1975, Paper No. 15, Special Publication No. 47, pp. 325–341

CEMENT GUN COMPANY, *Gunite chimney linings*. Technical Data Sheet 2, Published by the Company, London

GEORGE, C. M., *The Structural Use of High Alumina Cement Concrete*, Lafarge Fondu International, France, 1974, p. 15

PINFOLD, G. M., *Reinforced Concrete Chimneys and Towers*. The Cement and Concrete Association, London, 12.064, 1975, p. 233

APPLETON, B., Coming apart at the seals. *New Civil Engineer*, 1973, 6 December, pp. 6–9

PERKINS, P. H., *The protection of Portland cement concrete against sulphuric acid formed by bacterial action*. Paper at the Symposium held in London on 29 April 1976 by the Microbiology Division of the Institute of Petroleum

VARIOUS, Thirty-two papers, preprints of Conference held in London 19–21 January 1976 on *Underground Engineering—The Next Decade*, Foundation Publications Ltd., Leamington Spa

PERKINS, P. H., *Floors, Construction and Finishes*, Cement and Concrete Association, London, 1974, p. 132

AMERICAN CONCRETE INSTITUTE, Guide for the protection of concrete against chemical attack. Report of ACI Committee No. 515, *Journal A.C.I.*, December 1966

RICHARDS, J. D., The effect of various sulphate solutions on the strength and other properties of cement mortars at temperatures up to 80 °C. *Mag. Conc. Res.* (51) June 1965, pp. 69–76

APPENDIX 1

National Standards

A1.1 UK STANDARDS AND CODES OF PRACTICE

Cements	
Portland cement, ordinary and rapid hardening	BS 12
Portland cement, white and coloured	BS 12
Portland cement, sulphate-resisting	BS 4027
Portland cement, ultra-high early strength	None (Agrément Certificate 73/170)
Portland blastfurnace cement	BS 146
Portland blastfurnace cement, low heat	BS 4246
Portland cement, low heat	BS 1370
High alumina cement	BS 915
Supersulphated cement	BS 4284
Pozzolanic cement	None
Pigments for cement and concrete	BS 1014
Pulverised-fuel ash (pfa) for use in concrete	BS 3892
Cement Standards of the World, published by Cembureau, Paris, 1968	
Aggregates	
Aggregates from natural sources for concrete (including granolithic)	BS 882 and 1201
Building sands from natural sources	BS 1198–1200
Testing Concrete	
Methods of testing concrete	BS 1881
Methods for sampling and testing of mineral aggregates, sands and fillers	BS 812
Methods of sampling and testing lightweight aggregates for concrete	BS 3681
Tests for water for making concrete	BS 3148

Test sieves	BS 410
Ready-mixed concrete	BS 1926
Methods of specifying concrete	BS 5328
Recommendations for non-destructive methods of test for concrete	BS 4408

Steel Reinforcement

Expanded metal (steel) for general purposes	BS 405
Hard drawn mild steel wire for the reinforcement of concrete	BS 4482
Hot rolled steel bars for the reinforcement of concrete	BS 4449
Cold worked steel bars for the reinforcement of concrete	BS 4461
Steel fabric for the reinforcement of concrete	BS 4483
Steel wire for prestressed concrete	BS 2691
Stress relieved-wire steel strand for prestressed concrete	BS 3617
Galvanised steel reinforcement	None
Structural steel sections	BS 4
Galvanised iron and steel wire for telegraph and telephone purposes	BS 182–4
Galvanised wire netting	BS 1485
The use of structural steel in building	BS 449
Wrought steels—blooms, billets, bars & forgings	BS 970
(stainless, heat-resisting and valve steels)	BS 970 Part 4
Steel plate, sheet and strip	BS 1449
(stainless heat-resisting plate, sheet and strip)	BS 1449 Part 4
Hot-dip galvanised coatings for iron and steel	BS 729

Precast and Preformed Materials

Precast concrete flags	BS 368
Acid-resisting bricks and tiles	BS 3679

Joint Sealants

Cold poured joint sealants for concrete pavements	BS 5212
Hot applied joint sealing compounds for concrete pavements	BS 2499
Corrugated copper jointing strip for expansion joints for use in general building construction	BS 1878
Two-part polysulphide-based sealing compounds for the building industry	BS 4254
One-part gun-grade polysulphide-based sealants	BS 5215

Mastic asphalt for tanking and damp-proof courses	BS 1097 & 1418
Mastic asphalt for roofing (natural rock asphalt aggregate)	BS 988 & 1162
Water bars, rubber	None
Water bars, PVC	None
Neoprene joint sealants	None
Mastic asphalt for flooring	BS 1076, 1410 & 1451
Low and medium density polyethylene sheet for general purposes	BS 3012
Materials for damp proof courses	BS 743
Hot applied damp resisting coatings for solums	BS 2832
Black pitch mastic flooring	BS 1450 & 3672
Methods of test for fluidising plastics	BS 3712

General

Conversion factors and tables	BS 350
Bending dimensions and schedules of bars for the reinforcement of concrete	BS 4466
Glossary of terms for concrete and reinforced concrete	BS 2787
Glossary of general building terms	BS 3589
Fire tests on building materials and structures	BS 476
Building drawing practice	BS 1192
Schedule of weights of building materials	BS 648
Surface finish of blast-cleaned steel for painting	BS 4232
External rendered finishes (previously CP 221)	BS 5262
The structural use of concrete for retaining aqueous liquids	BS 5337
Basic data for the design of buildings: the control of condensation in dwellings	BS 5250
Code of basic data for the design of buildings	CP 3:
Thermal insulation in relation to the control of environment	Chapter II
Sound insulation and noise reduction	Chapter III
Precautions against fire	Chapter IV
Dead and imposed loads and wind loads	Chapter V
Engineering and utility services	Chapter VII
Heating and thermal insulation	Chapter VIII
Demolition	CP 94
Metal scaffolding	CP 97
Fire protection for electronic data processing installations	CP 95
Preservative treatments for constructional timber	CP 98

Foundations and substructures for non-industrial buildings of not more than four storeys	CP 101
Protection of buildings against water from the ground	CP 102
Structural recommendations for load-bearing walls	CP 111
The structural use of timber	CP 112
The structural use of concrete	CP 110
Structural use of reinforced concrete in buildings	CP 114
The structural use of prestressed concrete in buildings	CP 115
The structural use of precast concrete	CP 116
Composite construction in structural steel and concrete	CP 117
Dense concrete walls	CP 123, 101
Precast concrete cladding	CP 297
Natural stone cladding	CP 298
The protection of structures against lightning	CP 326
Sprayed asbestos insulation	CP 299
The design and construction of ducts for services	CP 413
Site investigations	CP 2001
Electric fire alarms	CP 327.404/402.501
Fire fighting installations and equipment	CP 402
Foundations	CP 2004
Protection of iron and steel structures from corrosion	CP 2008
Lining of vessels and equipment for chemical purposes	CP 3003 Parts 1–10
Cathodic protection	CP 1021
Sewerage	CP 2005
The structural use of concrete for retaining aqueous liquids	CP 2007 (now BS 5337)

(Note: following a recent decision by the British Standards Institution, Codes of Practice as a separate series will be phased out. All new Codes and revisions to existing Codes will be given a British Standard No., *i.e.* BS . . . .)

A1.2 SELECTED LIST OF RELEVANT ASTM STANDARDS (UNITED STATES OF AMERICA) RELATING TO CONCRETE

Cements	
Spec. for Portland cement	C 150–68
Spec. for air entraining Portland cement	C 175–68

Spec. for blended hydraulic cement	C 595–68
Chemical analysis of Portland cement	C 114–67
Test for fineness of Portland cement by the turbidimeter	C 115–67
Test for fineness of Portland cement by air permeability apparatus	C 204–55 (1967)
Test for specific gravity of hydraulic cement	C 188–44 (1967)
Test for heat of hydration of Portland cement	C 186–68
Test for normal consistency of hydraulic cement	C 187–68
Test for time of setting of hydraulic cement by Vicat needles	C 191–65
Test for time of setting of hydraulic cement by Gillmore needles	C 266–65
Test for compressive strength of hydraulic cement mortars	C 109–64
Test for autoclave expansion of Portland cement	C 151–66
Test for chemical resistance of mortars	C 267–65
Test for potential expansion of Portland cement mortars exposed to sulfate	C 452–68
Fly-ash and raw or calcined natural pozzolans for use in Portland cement concrete	C 618–68
Spec. for sieves for testing purposes (wire cloth sieves, round-hole and square-hole plate screens or sieves)	E 11–61 E 323–67

Aggregates

Spec. for concrete aggregates	C 33–67
Descriptive nomenclature of constituents of natural mineral aggregates	C 294–67
Rec. practice for petrographic examination of aggregates for concrete	C 295–65
Spec. for lightweight aggregates for structural concrete	C 330–68
Spec. for lightweight aggregates for concrete masonry units	C 331–64T*
Spec. for lightweight aggregates for insulating concrete	C 332–66
Test for sieve or screen analysis of fine and coarse aggregates	C 136–67
Test for materials finer than No. 200 sieve in mineral aggregates by washing	C 117–67

* T denotes Tentative Standard.

Test for specific gravity and absorption of coarse aggregate	C 127–68
Test for specific gravity and absorption of fine aggregate	C 128–68
Test for unit weight of aggregate	C 29–68
Test for voids in aggregate for concrete	C 30–37 (1964)
Test for surface moisture in fine aggregate	C 70–66
Test for organic impurities in sands for concrete	C 40–66
Test for clay lumps in natural aggregates	C 142–67
Test for lightweight pieces in aggregate	C 123–66
Test for soundness of aggregates by use of sodium sulfate or magnesium sulfate	C 88–63
Test for resistance to abrasion of small size coarse aggregate by use of the Los Angeles machine	C 131–66
Test for abrasion of rock by use of the Deval machine	D 2–33 (1968)
Test for abrasion of graded coarse aggregate by use of the Deval machine	D 289–63
Test for potential reactivity of aggregates (chemical method)	C 289–66
Test for potential alkali reactivity of cement–aggregate combinations (mortar bar method)	C 227–67
Test for potential volume change of cement–aggregate combinations	C 342–67

Concrete

Test for flow of Portland cement concrete by use of the flow table	C 124–39 (1966)
Test for slump of Portland cement concrete	C 143–66
Test for ball penetration in fresh Portland cement concrete	C 360–63
Test for time of setting of concrete mixtures by penetration resistance	C 403–68
Test for bleeding of concrete	C 232–58 (1966)
Test for weight per cubic foot, yield, and air content (gravimetric) of concrete	C 138–63
Test for air content of freshly mixed concrete by the volumetric method	C 173–68
Test for air content of freshly mixed concrete by the pressure method	C 231–68
Test for single-use molds for forming 6-in by 12-in concrete compression test cylinders	C 470–67T

Test for compressive strength of molded concrete cylinders	C 39–68
Making and curing concrete compression and flexure test specimens in the field	C 31–66
Test for compressive strength of concrete using portions of beams broken in flexure (modified cube method)	C 116–68
Making and curing concrete compression and flexure test specimens in the laboratory	C 192–68
Test for flexural strength of concrete (using simple beam with third-point loading)	C 78–64
Test for flexural strength of concrete (using simple beam with centre-point loading)	C 293–68
Obtaining and testing drilled cores and sawed beams of concrete	C 42–68
Test for static Young's modulus of elasticity and Poisson's ratio in compression of cylindrical concrete specimens	C 469–65
Test for fundamental transverse, longitudinal, and torsional frequencies of concrete specimens	C 215–60
Test for abrasion resistance of concrete	C 418–68
Test for length change of cement mortar and concrete	C 157–66T
Test for cement content of hardened Portland cement concrete	C 85–66
Spec. for air entraining admixture for concrete	C 260–66T
Rec. practice for microscopical determination of air-void content, specific surface, and spacing factor of the air-void system in hardened concrete	C 457–67T
Test for resistance of concrete specimens to rapid freezing and thawing in water	C 290–67
Test for resistance of concrete specimens to rapid freezing in air and thawing in water	C 291–67
Spec. for ready-mixed concrete	C 94–68

A1.3 CODES OF PRACTICE RELATING TO CONCRETE IN USE IN THE UNITED STATES OF AMERICA

American Concrete Institute

*Specifications for structural concrete for buildings	(ACI 302.66)

* Approved by USA Standards Institute as a USA standard.

*Building code requirements for reinforced concrete	(ACI 318–71)
Commentary on building code requirements for reinforced concrete	
Structural plain concrete	(322.1–322.9)
Tentative recommendations for design of concrete beams and girders for buildings	(333.1–20)
Concrete shell structures. Practice and commentary	(334.1–18)
Suggested design procedures for combined footings and mats	(336.1–16)
Recommended practice for shotcrete	(ACI 506–66)

A1.4 CANADA

CSA A23.1	Concrete, materials and methods of construction
A23.2	Methods of test for concrete
CSA A 5	Portland cement
CSA A 8	Masonry cement
19 GP–3b	Sealing compounds—two-part polysulphide, chemically curing
19 GP–5b	Sealing compounds—one-component silicone based, chemically curing
19 GP–13b	Sealing compounds—polysulphide one-component, chemically curing
CG SB 19GP–23	Standard for guide to selection of sealants on a use basis

A1.5 NATIONAL STANDARDS AND CODES: AUSTRALIA

For complete list of Publications and Subject Index of Australian Standards, 1976, reference should be made to the Standards Association of Australia, Standards House, 80–86 Arthur Street, North Sydney, N.S.W.

The following are some of the principal standards relating to cement and concrete:

A S 2	Portland cement
A S 1316	Masonry cement
A S 1317	Blended cement
A S 100 *et seq.*	Methods of testing concrete

* Approved by USA Standards Institute as a USA standard.

A S 1480 Concrete structures code
A S 1481 Prestressed concrete code
A S MP 27 Manual of physical testing of Portland cement
Admixtures: A173, CA.58, MP.20, 1478–9
Aggregates for concrete: 1465 to 1467
Tests on aggregates: 1141
Readymixed concrete: A64 1379
Reinforcement for concrete, bars, wire and rods: A81–84, A92, A97, 1302–1304

A1.6 NEW ZEALAND STANDARDS RELATING TO CEMENT AND CONCRETE

AGGREGATES
NZ3111: 1974 Methods of test for water and aggregates for concrete
NZS3121: 1974 Water and aggregate for concrete
NZS1958: 1965 Lightweight aggregate for structural concrete
NZS1959: 1965 Lightweight aggregate for concrete masonry units
NZS1960: 1965 Lightweight aggregates for insulating concrete

CEMENTS
NZS3122: 1974 Portland Cement (ordinary, rapid hardening, and modified)
NZS3123: 1974 Portland Pozzolan Cement

CONCRETE CONSTRUCTION
NZS3101P: 1970 Reinforced and plain concrete construction

CONCRETE MIXERS
NZS3105: 1975 Concrete Mixers

CONCRETE TESTING
NZS3112: 1974 Methods of test for concrete

FIRE RESISTANCE
NZ MP9 Reports on fire resistance ratings of elements of building structure
NZ MP9/1: 1962 Fire resistance ratings of walls and partitions: structures of concrete masonry blocks
NZ MP9/2: 1962 Fire resistance ratings of walls and partitions: structural concrete
NZ MP9/5: 1972 Fire resistance of prestressed concrete
NZ MP9/6: 1966 Fire resistance of beams and columns, excluding timber
NZ MP9/7: 1966 Fire resistance ratings of floor/ceiling combinations

FOUNDATIONS
NZ MP420400: 1973 Metric handbook to NZS4204P: 1973 Code of practice for foundations for buildings not requiring specific design
NZ MP420500: 1973 Metric handbook to NZS4205P: 1973 Code of practice for design of foundations for buildings
NZS4204P: 1973 Foundations
NZS4205P: 1973 Foundations

LIGHTWEIGHT CONCRETE (See also AGGREGATES)
NZ3152: 1974 Manufacture and use of structural and insulating lightweight concrete
NZS3141P: 1970 Precast lightweight concrete panels and slabs

MASONRY (See also 1900 below)
NZS3102P: 1974 Concrete Masonry units

PAVEMENTS
BS5215: 1975 Cold poured joint sealant for concrete pavements (endorsed by N.Z.)

PRECAST CONCRETE
NZ3151: 1974 Precast lightweight concrete panels and slabs

PRESTRESSED CONCRETE
NZSR 3R32: 1968 Prestressed concrete Amdt no. 1: 1970
NZ1417: 1971 Steel wire for prestressed concrete (Amended to suit NZ conditions from BS 2691: 1969)
NZ MP32x000: 1974 Metric handbook to NZSR32: 1968 Prestressed Concrete

READY MIXED CONCRETE
NZS2086: 1974 Ready mixed concrete production
MP208600: 1975 Metric handbook to NZS2086: 1968 Ready Mixed Concrete

REINFORCED CONCRETE
NZ3101P: 1970 Code of practice for reinforced concrete—design
NZ MP310100: 1973 Metric handbook to NZS3101P: 1970 Code of Practice for reinforced concrete design

REINFORCEMENT
NZS3402P: 1973 Hot rolled steel bars for concrete reinforcement
NZS3403 Hot-dip galvanised corrugated or profiled steel sheet
NZS3421: 1975 Hard drawn mild steel wire for concrete reinforcement

NZS3441 Hot-dip galvanised plain steel sheet and strip (to be metricated late 1976)
NZS3422: 1975 Welded fabric of drawn steel wire for concrete reinforcement

SANDS
NZ2129 Sands for mortars, plasters and external renderings Amdt. no. 1, 1971 (in revision May 1976)

SHELLS
NZ1826: 1964 The design and construction of shell roofs

SWIMMING POOLS
NZS4441: 1972 Code of Practice for swimming pools

1900 (MODEL BUILDING BYLAW)
Ch 6.2: 1964 Masonry Amdt. No. 1, 1973 (revision in progress May 1976)
Ch 9.2: 1964 Design and construction Amdt No. 1, 1965, No. 2, 1973 (revision in progress May 1976) (to be reconstituted incorporating NZS2086)
Ch 9.3A: 1970 Concrete—general requirements and materials and workmanship no. 11970 n2 1971, no. 3, 1973 (revision in progress May 1976) (to be reconstituted incorporating NZS2086)
Ch. 11. Special structures
11.1: 1964 Concrete Structures for the storage of liquids Amdts no. 1, 1964, no. 2, 1973 (general revision in progress May 1976)
MP 190093A: Metric handbook to NZS1900 Chapter 9.3A: 1970 Concrete Design and construction general requirements

A1.7 INDIA

IS 476 Code of Practice for plain and reinforced concrete
IS 269 Portland cement, ordinary rapid hardening and low heat
IS 455 Portland blastfurnace cement
IS 1489 Pozzolanic cement
IS 516 Methods of test for strength of concrete
IS 3466 Masonry cement
IS 6452 High alumina cement for structural use

A1.8 SOUTH AFRICA: SABS STANDARDS AND CODES OF PRACTICE

STANDARDS (METRIC UNITS, UNLESS OTHERWISE INDICATED)
28–1972 Metal ties for cavity walls
82–1976 Bending dimensions of bars for concrete reinforcement

197–1971	Test sieves
248–1973	Bituminous damp-proof courses
471–1971	Portland cement and rapid-hardening portland cement
523–1972	Limes for use in building
527–1972	Concrete building blocks
541–1971	Precast concrete paving slabs
626–1971	Portland blastfurnace cement
794–1973	Aggregates of low density
831–1971	Portland cement 15 and rapid-hardening cement 15
878–1970	Ready mixed concrete
920–1969	Steel bars for concrete reinforcement
927–1969	Precast concrete kerbs and channels (*not metric*)
973–1970	Standard form of specification for concrete work
975–1970	Prestressed concrete pipes
986–1970	Precast reinforced concrete culverts
987–1970	Cement bricks
993–1972	Modular co-ordination in building
1024–1974	Welded steel fabric for concrete reinforcement
1083–1976	Aggregates from natural sources

CODES OF PRACTICE

03A–1969	The protection of dwelling houses against lightning
021–1973	Waterproofing of buildings
043–1965	The laying of wood floors
058–1955	Sewer and drain jointing (*not metric*)
062–1956	Fixing of concrete roofing tiles (*not metric*)
073–1974	Safe application of masonry-type facings to buildings
088–1972	Pile foundations
0102–1968	The structural design and installation of precast concrete pipelines
0109–1969	Floor finishes on concrete

STANDARD BUILDING REGULATIONS

Note: The Standard Building Regulations cover all aspects of building construction in areas controlled by local authorities, although not all cities and towns in South Africa have adopted them. The main chapter headings cover definitions, administration in the four provinces of South Africa, loads and forces, foundations, plain and reinforced concrete, structural steelwork, structural timber, masonry and walling, miscellaneous materials and construction, water supply, lighting, drainage and sewerage, ventilation, fire protection, public safety, urban aesthetics and advertising.

DEPARTMENT OF INDUSTRIES

No. R.1830.] [23rd October, 1970.

STANDARDS ACT, 1962

STANDARD BUILDING REGULATIONS

In terms of section 14*bis* (1) of the Standards Act, 1962 (Act No. 33 of 1962) the Council of the South African Bureau of Standards, with the approval of the Minister of Economic Affairs, hereby publishes standard building regulations under the chapter headings listed in this notice.

LIST OF CHAPTERS

Chapter
1 DEFINITIONS AND INTERPRETATION
2 ADMINISTRATION, PART I, FOR THE PROVINCE OF THE CAPE OF GOOD HOPE
ADMINISTRATION, PART II, FOR THE PROVINCE OF THE TRANSVAAL
ADMINISTRATION, PART III, FOR THE PROVINCE OF NATAL
ADMINISTRATION, PART IV, FOR THE PROVINCE OF THE ORANGE FREE STATE
3 LOADS AND FORCES
4 FOUNDATIONS
5 PLAIN AND REINFORCED CONCRETE
6 STRUCTURAL STEELWORK
7 STRUCTURAL TIMBER
8 MASONRY AND WALLING
9 MISCELLANEOUS MATERIALS AND CONSTRUCTION
10 WATER SUPPLY
11 LIGHTING
12 DRAINAGE AND SEWERAGE
13 VENTILATION
14 FIRE PROTECTION
15 PUBLIC SAFETY
16 URBAN AESTHETICS
17 ADVERTISING

A1.9 WEST GERMAN STANDARDS (DINs)

Concrete and reinforced concrete structures; design and construction DIN 1045

Reinforcing steel; definitions, properties, markings	DIN 488
Testing methods for concrete	DIN 1048
Steel in structural engineering; design and construction	DIN 1050
Masonry; design and construction	DIN 1053
Design loads for buildings	DIN 1055
Highway bridges; design loads	DIN 1072
Steel highway bridges; design principles	DIN 1073
Composite girder highway bridges; code of practice for design and analysis	DIN 1078
Symbols for structural design calculations in civil and structural engineering	DIN 1080
Quality control in concrete and reinforced concrete construction	DIN 1084
Portland cement, 'Eisen' portland cement, 'Hochofen' cement and trass cement; definitions, constituents, requirements, delivery	DIN 1164
Reinforced lightweight concrete slabs	DIN 4028
Assessment of waters, soils and gases aggressive to concrete	DIN 4030
Water pressure retaining bituminous seals for structures; code of practice for design and construction	DIN 4031
Welding of reinforcing steel; requirements and tests	DIN 4099
Fire behaviour of materials and components of structures; definitions, requirements and tests for components	DIN 4102
Light partitions; code of practice for construction	DIN 4103
Thermal insulation in building construction	DIN 4108
Sealing of structures against ground moisture; code of practice for construction	DIN 4117
Intermediate components of concrete for reinforced and prestressed concrete floors	DIN 4158
Floor blocks and blocks for wall panels, structurally co-operating	DIN 4159
Contract procedure for building work, Part C, technical specification for grouting work	DIN 18 309
Contract procedure for building work, general technical specification for concrete and reinforced concrete work	331
Contract procedure for building work, general technical specification for roof coverings and sealing work	338
Contract procedure for building work, general technical specification for floor screeding work	353
Contract procedure for building work, general technical specification for flooring work	365

Contract procedure for building work, general technical specification for damp-proofing against water under pressure	336
Provisional. Air-placed concrete, production and tests	DIN 18 551
Contract procedure for building work. Part B, General contract conditions for execution of building work	DIN 1961
Concrete for sewage works, manufacture and testing	DIN 4281
Floor blocks, not structurally co-operating	DIN 4160
Perforated plates for test screens; square holes	DIN 4178
Sieve bottoms; wire-mesh sieves for testing dimensions	DIN 4188
Slag-based cementitious materials ('Mischbinder')	DIN 4207
Aids for the design and analysis of reinforced concrete components	DIN 4224
Aggregates for concrete	DIN 4226
Loadbearing walls of no-fines lightweight concrete; design and construction	DIN 4232
Internal vibrators for the compaction of concrete; code of practice for their use	DIN 4235
Vibrating tables for the compaction of concrete; code of practice for their use	DIN 4236
Composite beams—building construction; code of practice for design and construction	DIN 4239
Glass filler elements for concrete requirements, testing	DIN 4243
Regulations for scaffolding	DIN 4420
Mortars made with mineral binders; testing	DIN 18 555
Testing of natural stone; code of practice for testing and selecting natural stone	DIN 52 100

Portland Cements

Type	Name	Grade	Standard
OCI	Portlandzement	Z 375 I	DIN 1164 1969
OCII		Z 375 (f)	
HSC I		Z 450 I	
HSC II		Z 450 f	
HSC III		Z 530	

(Note: The figures '375–450' etc. refer to 28 day compressive cube strength expressed in kg/cm^2)

A1.10 AUSTRIA

A 2050	Building contracts, placing contracts
ONORM B 2110	Building construction, general conditions for building contracts

ONORM B 2211 Concrete and reinforced concrete construction
ONORM B 3303 Testing concrete
ONORM B 3305 Water, soils and gases aggressive to concrete: evaluation and chemical analysis
ONORM B 4200 Concrete structures, design and construction
ONORM B 4250 Prestressed concrete structures
ONORM B 3310 Portland cements, Types OCI, OCII, HSC, and blastfurnace cements
ONORM B 4200 Part 7 Steel reinforcement for concrete

A1.11 NETHERLANDS

NEN 3861 & 3862 Regulations for concrete (these partly supersede NEN 1009)
NEN 6008 Steel reinforcement for concrete
N 481 Portland cement
N 483 and 484 Portland blastfurnace cements
N 618 Pozzolanic cement

A1.12 ITALY

UNI 595 Portland cement, blastfurnace cement and pozzolanic cements
UNI 6132 & 6135 Testing of concrete

A1.13 BELGIUM

NBN 15 Concrete construction
NBN 15 201–205, 211–214, 216, 218, 220, 227, and 250. Concrete tests
NBN A24.301–303, Steel reinforcement for concrete
NBN 460.01–03 Wind loads on structures
NBN 48 Portland cement
NBN 130, 131, and 198. Portland blastfurnace cement

A1.14 FRANCE

CC BA 68 Reinforced concrete Code
CM 66 Structural steelwork Code

NV 65 Snow and wind loading
PO 3–011 Standard contracts for private building work, general clauses
NF P 15–302 Portland cement
 303, 304, 305 & 311 Portland blastfurnace cement
NF P 18–102, 404, 405, 406 Tests for concrete
NF A.35.015 and 016. Steel reinforcement for concrete

APPENDIX 2

Conversion Factors and Coefficients

$1\ m^2 = 10{\cdot}7\ ft^2$
$1\ ft^2 = 0{\cdot}093\ m^2$
$1\ kg = 2{\cdot}26\ lb$
$1\ lb = 0{\cdot}45\ kg$
$1\ m = 3{\cdot}28\ ft$
$1\ ft = 0{\cdot}31\ m$
$1\ lbf/ft^2 = 4{\cdot}85\ kg/m^2$

Conversion from Imperial to metric

Di = Density in lb/ft^3
Dm = Density in kg/m^3

$$Dm = \frac{Di}{62{\cdot}4} \times \frac{1000}{1} = 16\,Di\,(approx.)$$

Density of structural concrete, made with aggregates from natural sources

$148\ lb/ft^3 = 4000\ lb/yd^3 = 2385\ kg/m^3 = 2400\ kg/m^3$ (approx.)

Bulk densities of concreting materials

Cement	$1400\ kg/m^3$	$= 88\ lb/ft^3$
Sand	$1600\ kg/m^3$	$= 100\ lb/ft^3$
Coarse aggregate	$1450\ kg/m^3$	$= 91\ lb/ft^3$

These figures are very approximate.

Modulus of elasticity of concrete: $E = 4{\cdot}5 \times 10^6$.
Range: $E = 2{\cdot}5$ to 6×10^6.

Specific gravity

water	1
cement, Portland	3·12
cement, high alumina	3·20
pit sand	2·65
flint gravel	2·55
limestone	2·80
granite	2·75
basalt (Whinstone)	2·90
concrete structural	2·38

Specific heat

	Imperial	Metric
concrete	0·25 Btu	0·063 k/cal
mild steel	0·12 Btu	0·031 k/cal

Latent heat

fusion of ice	144 Btu/lb	37 k/cal/gm
evaporation of water	970 Btu/lb	250 k/cal/gm

1 Btu/lb = 2326 J/kg

Force

1 lbf = 4·45 N
1000 lbf/in^2 = 7 N/mm^2 (approximately)
143 lbf/in^2 = 1 N/mm^2 (approximately)
1 kN/mm^2 = 1000 N/mm^2
1 MN/m^2 = $10^6 N/m^2$ = 1 N/mm^2

Thermal transmittance (U) value

$$\text{Btu/ft}^2\text{ hour }^\circ\text{F} = \frac{\text{W/m}^2\ ^\circ\text{C}}{5{\cdot}678}$$

$$= 0{\cdot}176\,(\text{W/m}^2\ ^\circ\text{C})$$

$$\text{W/m}^2\ ^\circ\text{C} = 5{\cdot}678\,(\text{Btu/ft}^2\text{ hour }^\circ\text{F})$$

Thermal conductivity, coefficient of heat transfer

1 Btu ft/ft^2 hour °F = 1·731 W/m °C

Coefficient of thermal expansion of concrete

Imperial: Range: $3{\cdot}5 \times 10^{-6}$ to $6{\cdot}5 \times 10^{-6}$ °F
Metric: Range: $6{\cdot}3 \times 10^{-6}$ to $11{\cdot}7 \times 10^{-6}$ °C

To convert lb/yd^3 of compacted concrete to kg/m^3 multiply by 0·6.

To convert kg/m^3 of compacted concrete to lb/yd^3 multiply by 1·67.

Concrete strengths are designated in Newtons per square millimetre (N/mm^2). Sometimes mega-Newtons per square metre (MN/m^2) are used. The two units are equal numerically.

APPENDIX 3

Construction Associations and Research Organisations

A3.1 UNITED KINGDOM

(1) The Cement and Concrete Association
52 Grosvenor Gardens
LONDON SW1W 0AQ
Telephone: 01-235 6661

(2) The Building Research Establishment
Garston
WATFORD, Herts
Telephone: Garston (092 73) 76612 and 01-477 4040

(3) The British Standards Institution
2 Park Street
LONDON W1A 2BS
Telephone: 01-629 9000

(4) The British Concrete Pumping Association
c/o Kitsons Taylor & Co
52 Lincoln's Inn Fields
LONDON WC2
Telephone: 01-405 9292

(5) British Precast Concrete Federation
60 Charles Street
LEICESTER LE1 1FB
Telephone: Leicester (0533) 28627

(6) British Ready-Mixed Concrete Association
Shepperton House
Green Lane
SHEPPERTON
Middlesex
Telephone: Walton-on-Thames 43232

(7) The Aluminium Federation
60 Calthorpe Road
Fiveways
BIRMINGHAM 15
Telephone: 021-454 3805

(8) The Construction Industry Research and Information Association
6 Storey's Gate
LONDON SW1
Telephone: 01-839 6881

(9) The British Constructional Steel Association (CONSTRADO)
12 Addiscombe Road
CROYDON CR9 3JH
Telephone: 01-686 0366

(10) The Mastic Asphalt Council
24 Grosvenor Gardens
LONDON SW1W 0DH
Telephone: 01-730 7175

(11) Reinforcement Manufacturers Association
16 Tooks Court
LONDON EC4
Telephone: 01-242 4259

(12) The Stainless Steel Development Association
65 Vincent Square
LONDON SW1
Telephone: 01-834 6737

(13) The Steel Sheet Information and Development Association
Albany House
Petty France
LONDON SW1
Telephone: 01-799 1616

(14) The Corrosion Advice Bureau
Corporate Engineering Laboratory
British Steel Corporation
140, Battersea Park Road
LONDON SW11 4LZ
Telephone: 01-622 5511

(15) The British Ceramic Tile Council
Federation House
STOKE-ON-TRENT
Staffs
Telephone: Stoke-on-Trent (0782) 45147

(16) The Brick Development Association
19 Grafton Street
LONDON W1X 3LE
Telephone: 01-409 1021

A3.2 WEST GERMANY

Bundesverband der Deutschen Zementindustrie e.V.
Riehlerstrasse 8
(Postfach 140105)
5 Köln 1
T 73 00 76 C Zementverband
Tx 08881603
(Cement industry and technical advisory services)

Forschungsinstitut der Zementindustrie
Tannenstrasse 2
4 Düsseldorf-Nord
T 43 44 51 C Zementforschung
Tx 08584876
(Research Institute of the cement industry)

Verein Deutscher Zementwerke e.V.
Tannenstrasse 2
4 Düsseldorf-Nord
T 43 44 51 C Zementforschung
Tx 08584876
(Technical association of the cement industry)

Deutscher Beton-Verein e.V.
Bahnhofstr 61
62 Wiesbaden
(Concrete)

A3.3 UNITED STATES OF AMERICA

Portland Cement Association
Old Orchard Road
Skokie
Illinois
T (312) 9666200

(Research, information and technical advisory services)

American Concrete Institute
P.O. Box 4754
Redford Station
Detroit, Michigan 48219

(Technical and Educational Committee activities and publications for manufacturers and users of concrete)

Architectural Precast Association (APA)
2201 East 46th Street
Indianapolis, Indiana 46205
T (317) 2511214

(Precast concrete)

National Concrete Masonry Association (NCMA)
Pompino East Building–1800 Kent Street
P.O. Box 9185 Rosslyn Station
Arlington, Virginia 22209
T (703) 5240815

(Precast concrete)

National Precast Concrete Association (NPCA)
2201 East 46th Street
Indianapolis, Indiana 46205
T (317) 2530486

(Precast concrete)

Prestressed Concrete Institute (PCI)
20 North Wacker Drive
Chicago, Illinois 60606
T (312) 3464071

(Prestressed concrete)

American Concrete Pipe Association (ACPA)
1501 Wilson Boulevard
Arlington, Virginia 22209
T (703) 5243939

(Concrete pipes)

American Concrete Pressure Pipe Association (ACPPA)
1501 Wilson Boulevard
Arlington, Virginia 22209
T (703) 5243939
(Concrete pipes)

Concrete Reinforcing Steel Institute (CRSI)
228 North LaSalle Street
Chicago, Illinois 60601
T (312) 3725059
(Concrete reinforcing steel)

National Ready Mixed Concrete Association (NRMCA)
900 Spring Street
Silver Spring, Maryland 20910
T (310) 5871400
(Ready mixed concrete)

Expanded Shale, Clay and Slate Institute
1041 National Press Building
Washington, D.C. 20004
(Structural lightweight concrete)

A3.4 CANADA

Portland Cement Association
116 Albert Street
Ottawa
T 236 9471
(Research, information and technical advisory services)

Canadian Prestressed Concrete Institute (CPCI)
120 Eglington Avenue, East
Toronto 12
Ontario
T (416) 4895616
(Prestressed concrete)

National Concrete Producers' Association (NCPA)
3500 Dufferin Street
Suite 101
Downsview 460
Ontario
T (416) 6301204
(Concrete industry)

A3.5 AUSTRALIA

Cement and Concrete Association of Australia
147–152 Walker Street
North Sydney, N.S.W. 2060

Concrete Institute of Australia (Concrete research)
147 Walker Street
North Sydney, N.S.W. 2060
T 92 0316

Concrete Masonry Association of Australia (Concrete masonry)
147 Walker Street
North Sydney, N.S.W. 2060
T 92 0316

National Ready Mixed Concrete Association (Ready mixed concrete)
332 Albert Street
Melbourne, Victoria 3001
T 419 1313

Precast Concrete Manufacturers Association (Precast concrete)
12 O'Connell Street
Sydney, 2001
T 250 5401

C.S.I.R.O. Division of Building Research
P.O. Box 56
Highett, Victoria 3190

Department of Construction, Experimental Building Station
P.O. Box 30
Chatswood, N.S.W. 2067

It should be noted that most Universities carry out research on concrete.

A3.6 NEW ZEALAND

N.Z. Portland Cement Association Box 2792 *Wellington*	(Information and technical advisory services)
N.Z. Concrete Masonry Association Box 9130 *Wellington*	(Concrete masonry)
N.Z. Concrete Products Association Box 9130 *Wellington*	(Concrete products)
N.Z. Prestressed Concrete Institute 11th floor Securities House 126 The Terrace P.O. Box 969 *Wellington C.*1.	(Prestressed concrete)
N.Z. Ready Mixed Concrete Association Box 12013 *Wellington*	(Ready mixed concrete)

NZ CONCRETE RESEARCH & INFORMATION ORGANISATIONS
Associations & Related Bodies
Building Research Association of New Zealand,
42 Vivian Street,
P.O. Box 9375 (Postal),
Wellington, N.Z.

Department of Scientific and Industrial Research,
Chemistry Division,
Gracefield Road,
Lower Hutt, N.Z.,
Private Bag (Postal),
Petone, N.Z.

Department of Scientific and Industrial Research,
Physics and Engineering Laboratory,
Gracefield Road,
Lower Hutt, N.Z.,
Private Bag (Postal),
Lower Hutt, N.Z.

Ministry of Works and Development,
Vogel Building,
Aitken Street,
P.O. Box 12-041 (Postal),
Wellington 1, *N.Z.*

N.Z. Concrete Research Association,
13 Wall Place,
Tawa, N.Z.,
P.O. Box 50-156 (Postal),
Porirua, N.Z.

N.Z. Institute of Architects,
Maritime Building,
2–10 Customhouse Quay,
P.O. Box 438 (Postal),
Wellington 1, *N.Z.*

N.Z. Institution of Engineers,
Molesworth House,
101 Molesworth Street,
P.O. Box 12-241 (Postal),
Wellington 1, *N.Z.*

University of Auckland,
School of Engineering,
24 Symonds Street,
Private Bag (Postal),
Auckland, N.Z.

University of Canterbury,
School of Engineering,
Creyke Road,
Ilam,
Christchurch 4,
Private Bag (Postal),
Christchurch, N.Z.

A3.7 REPUBLIC OF SOUTH AFRICA

South African Cement Producers Association (Cement industry)
P.O. Box 61514
Marshalltown
T 8387502 C safcem

National Building Research Institute (Research, information and technical advisory services)
P.O. Box 395
Pretoria
T 74 6011 C navorsbou

PORTLAND CEMENT INSTITUTE

Head Office:
18 Kew Road
Richmond
Johannesburg
2092

Natal Regional Office:
P.O. Box 90
Westville
Natal
3630

Western Cape Regional Office:
Molteno Street
Goodwood
Cape Province
7460

Eastern Cape Regional Office:
P.O. Box 1540
Port Elizabeth
Cape Province
6000

SOUTH AFRICAN COUNCIL FOR SCIENTIFIC & INDUSTRIAL RESEARCH
P.O. Box 395
Pretoria
Transvaal
0001
National Building Research Institute
National Institute for Transport and Road Research

RESEARCH (ONLY) ORGANISATIONS

South African Railways Research Laboratories
Leyds Street
Johannesburg
2001

UNIVERSITIES

The Department of Civil Engineering at the following Universities:

University of the Witwatersrand
1 Jan Smuts Avenue
Johannesburg
2001

(This university also has a department of Building Science)

University of Cape Town
Private Bag
Rondebosch
Cape Province
7700

University of Natal
King George V Avenue
Durban
4001

Randse Afrikaanse Universiteit
Posbus 524
Johannesburg
2000

University of Pretoria
Brooklyn
Pretoria
0181

University of Stellenbosch
Stellenbosch
Cape Province
7600

APPENDIX 4

Notes on the Testing of Concrete in Structures

There are three principal methods of testing the strength of concrete in a structure:

1. Non-destructive testing
   (a) The Schmidt rebound hammer
   (b) Gamma radiography
   (c) Ultrasonic pulse velocity

   To the above can be added the use of an electromagnetic cover meter for determining whether there is in fact steel reinforcement in the concrete, and if so its depth below the surface from which the instrument is used.
2. Core taking and testing

   This is only partly non-destructive as the core holes have to be made good, and in a water-retaining or water-excluding structure this is not easy.
3. Load testing on individual members or groups of members

Each of the three methods given above is a subject in its own right and readers who wish to obtain detailed information should refer to the Bibliography at the end of this appendix, as only a few general notes will be given here.

Methods in (1) are covered by a British Standard, BS 4408: Recommendations for Non-Destructive Methods of Test for Concrete. Method (2) is covered by BS 1881: Methods of Testing Concrete, and Method (3) is dealt with in Code of Practice CP 110: The Structural Use of Concrete, as well as the three 'parallel' Codes, CP 114, CP 115 and CP 116.

A4.1 NON-DESTRUCTIVE TESTING

A4.1.1 The Schmidt Rebound Hammer

This is a useful and practical instrument for site work, but the results must be interpreted with caution taking into account its limitations.

The instrument measures the surface hardness of the concrete, but experience based on tests has shown that this (the rebound number) can be related empirically to the compressive strength, provided certain precautions are taken. These are:

(i) the instrument should be calibrated on the type of concrete being tested; 150 mm cubes are better than 100 mm.
(ii) a minimum of 15 readings should be taken on each structural unit.
(iii) the instrument should be recalibrated when used on concrete made with cements other than Portland, *e.g.* HAC.
(iv) The highest and the lowest reading should be eliminated and the average taken of the remainder. The average rebound number is then referred to the calibration chart and the strength read off. Alternatively, each of the readings, except those rejected, can be converted to compressive strength, and then these strengths are averaged.

It is useful to compare the readings on concrete which has been accepted as satisfactory with similar concrete in similar structural unit(s) which has given cause for concern. The usual reason for this 'concern' is that the cube results were below the specified minimum. If the readings on the suspect unit(s) are significantly lower, then this would support the view that the concrete was in fact of lower compressive strength. However, the results are only indicative, and if it is decided to continue with the investigation, the next step would be to take cores. It is of course assumed that the sampling and testing procedure for the cubes has been checked and found to be in order.

British Standard, BS 4408: Part 4, gives recommendations for surface hardness methods of non-destructive tests for concrete.

A4.1.2 Gamma Radiography

This method of inspection of hardened concrete is suitable for thicknesses up to about 450 mm. It should be noted that gamma radiography comes within the scope of the Factories Acts and the Ionising Radiations (Sealed Sources) Regulations 1961. It is used for checking the efficiency of the grouting in cable ducts in post-tensioned prestressed concrete, to detect voids (honeycombing) in concrete, and to determine accurately the position and size of reinforcement.

Recommendations for this method of non-destructive testing are given in BS 4408: Part 3. The work is expensive and requires specialist operators as well as experience in the interpretation of the photographs; also it must be possible to have access to both sides of the structural unit under investigation.

A4.1.3 Ultrasonic Pulse Velocity Testing

The author wishes to record that the information which follows is largely based on discussions which he has had with Mr. H. S. Tomsett of the Cement and Concrete Association and a reference to a paper by Mr. Tomsett is given in the Bibliography at the end of this chapter.

The basic principle of ultrasonic testing of concrete in the structure is that concrete is an elastic material and will transmit longitudinal, compression, and shear waves. The velocity with which these waves travel through the concrete is determined by the properties of the concrete which control the elastic modulus. These properties are in turn related to the strength of the concrete. The apparatus used generates a pulse in the concrete by the application of a mechanical impulse; it collects the impulse at some point at a measured distance from the point of generation, and contains a timing mechanism which accurately measures the time taken for the leading edge of the pulse to pass from the transmitter to the receiver.

A number of advisory engineers in the Cement and Concrete Association have over the past few years used this ultrasonic pulse velocity equipment on numerous sites in various parts of the country. In this way they have gathered considerable experience of the use of the equipment and perhaps more important the practical interpretation of the results so that the quality of the concrete as based on strength can be reasonably assessed. The apparatus used is that developed by Elvery at University College London and produced by C.N.S. Instruments Limited. It has the trade name of Pundit (Portable Ultrasonic Non-destructive Digital Indicating Tester). The apparatus is small and readily portable as it weighs only about 3 kg.

There are three basic ways in which the pulse can be transmitted and recorded. There is direct transmission, which is the most satisfactory; in this case the time measured is that for the longitudinal compression wave to pass between the transmitter and the receiver; the transmitter and receiver are on opposite sides of the structural unit concerned. The next method, which is not so satisfactory as the previous one, is semidirect, and is used for such units as thick floor slabs where access can be obtained to the top surface of the unit and to the side, but not to the underside. The third method which is the least satisfactory, is indirect, where transmission and receiving have to be carried out from one side only as in ground floor slabs and basement walls after back-filling has been carried out.

There is a British Standard BS 4408: Recommendations for Non-Destructive Methods of Test for Concrete, of which Part 5: The Measurement of the Velocity of Ultrasonic Pulses in Concrete, was published in 1974.

In this book it is not possible to go into the details of ultrasonic pulse velocity testing but those who consider using it, and there is no doubt it is an extremely useful method of testing concrete in the structure, should bear in mind the factors which affect the pulse velocity through concrete; these have been listed by Tomsett as:

aggregate type
aggregate grading
cement type
cement content
w/c ratio
degree of compaction of the concrete
curing temperature

The presence of reinforcement will affect the velocity obtained, and it is therefore advisable to make every effort to transmit the pulses in the concrete structure between the reinforcing bars as far as this is practicable.

When used by an experienced engineer, the Pundit can in most cases satisfactorily determine the depth of cracks and the location and extent of undercompacted poor quality concrete.

The author feels that the ultrasonic pulse velocity method of non-destructive testing is particularly applicable to water-retaining structures because the taking of cores in such structures below the water line is something which should be avoided whenever possible.

A4.1.4 Cover Meter Surveys by Electromagnetic Cover Measuring Devices

This is a comparatively simple piece of equipment, and the cost of a reliable model, in 1976, was about £150. The cover to reinforcement is shown on a graduated scale, and on a construction site the accuracy of a good commercial instrument is about ± 5 mm.

It is normal to calibrate it for mild steel and for use with Portland cement concrete, but it can be used satisfactorily with other types of cement and high tensile steel, provided it is specially calibrated for the purpose. Some types of stainless steel are nonmagnetic and so this type of instrument will not detect the metal; in such cases the alternative is gamma radiography.

Change in bar diameter from say 10 mm to 32 mm will not appreciably affect the cover shown, but for small diameter bars of say 5 mm, there will be a considerable difference between the indicated cover and the true cover; the cover shown on the scale is likely to be greater than the actual cover.

The relevant British Standard is BS 4408: Part 1.

A4.2 CORE TESTING

Core testing of the concrete in a structure is generally the last but one resort when cube results are significantly below the specification and the concrete is seriously suspected of being below the required strength. It is expensive and time consuming, and the results may not be as clear cut as one would wish. The cost in 1975 of taking 150 mm cores through a 250 mm thick floor or roof slab can cost £40–£50 each, excluding testing.

The taking of cores in a liquid retaining structure should not be lightly undertaken, as it means boring holes 100 mm or 150 mm diameter through the structure and it is not easy to ensure a watertight joint afterwards when these are reinstated.

The cores should be taken and tested in accordance with BS 1881: Methods of Testing Concrete. The standard gives detailed recommendations for the preparation, testing and examination of cores, but the interpretation of results requires considerable experience.

If the coring, testing and examination is carefully carried out it will give a considerable amount of information about the quality of the concrete in the structure. On the basis of the core results, a decision can be taken as to whether the concrete is satisfactory or whether the section in question may have to be demolished, and this is a very serious matter. When the results are still not entirely clear, a load test may be ordered.

In some cases, attempts are made to establish a correlation between the core strengths and the cube strengths. While it may be possible to establish a reasonable relationship under strictly controlled conditions, the author does not consider that this can be done satisfactorily on normal construction sites. The cubes are usually taken at the point of delivery of the concrete from the ready-mixed concrete lorry, or near the site batching plant. The cubes are then made, cured and tested in accordance with BS 1881.

The history of the two lots of concrete (cubes and that in the formwork) is quite different. The cubes are dealt with in a detailed and meticulous way, while the bulk of the concrete is transported by skip or pump, dropped into formwork containing reinforcement, and compacted by poker vibrators by men working perhaps in strong winds and driving rain.

Apart from what has been said above, the following also introduce variables between cube and core strength:

- the presence of reinforcement in the core;
- the difference in age; cores are usually taken later than 28 days after casting the concrete; see also remarks below;
- depending on the position in the structure, the core may have dried out a little more on one side than on the other; a curing membrane would influence the drying out process;
- there may be a noticeable difference in compaction between one end of a core and the other, particularly in floor slabs; in cores taken at the bottom of a lift and near the top, as in walls and columns the strength of the former may be 20% higher than the latter.

In view of the above, the author considers it unrealistic to try and assess the quality of the concrete as it was delivered to site (or as it left the batching plant), on the basis of core test results. A core will only indicate the type and quality of the concrete in the structure.

A further complication is the rate of gain of strength with age between the concrete in the structure and the test cubes. Codes of Practice, standards, and specifications normally assume that the rate of gain of strength for both concretes (cubes and structure) is the same. As far as the author is aware, this vital assumption has only recently been queried. Work and investigations by W. Murphy of the Cement and Concrete Association have raised serious doubts about the validity of this age–strength relationship in structures. Conclusions to date suggest that in fact there may be little increase in strength in the concrete in what may be termed a normal building structure after about 28 days from casting. There is no doubt however, that test cubes, dealt with as prescribed in BS 1881 do gain in strength after the first 28 days. As a guide only, the strength of concrete in a structure can be taken as about 70% of the average cube strength.

A4.3 LOAD TESTING

Load tests are only likely to arise for suspended slabs and beams including supporting columns when all other methods of non-destructive testing which are practical to the circumstances of the case have failed to resolve satisfactorily the question as to whether or not the part of the structure in question is structurally sound.

This type of testing is covered by Section 9 of British Standard Code of Practice, CP 110: The Structural Use of Concrete. For readers who prefer to keep to the original Codes (CP 114, 115, and 116), the relevant clauses are: CP 114: Clause 605; CP 115: Clause 602; CP 116: Clauses 504 and 505.

A useful paper on the load testing of insitu reinforced concrete structures in accordance with CP 114, was presented by G. F. Blackledge at the Symposium on Concrete Quality, held in London in November 1964. In this paper the author drew attention to the considerable differences in the requirements of the British Code and the American Concrete Institute Standard 318-63 and the Australian Code for Concrete in Buildings AS-CA2-1963. Doubts as to the structural soundness of a structure or part of a structure can arise from a number of reasons, of which the following are the most usual:

failure of test cubes, followed by failure to resolve this by core testing and/or other non-destructive tests;
an acknowledged error in design;
an acknowledged or suspected error in construction;

cracking and/or excessive deflexion;
an increase in operating load conditions not anticipated during design and construction.

Load testing is a tedious business and should only be resorted to when all else has failed and the responsible engineer is faced with a decision to either order the demolition of the structure (or part) or to accept a structure of which he has doubts about its strength.

A4.4 BIBLIOGRAPHY

SHACKLOCK, B. W., *Concrete Constituents and Mix Proportions*, Cement and Concrete Association, London, 1974, p. 102

MURDOCK, L. J. AND BLACKLEDGE, G. F., *Concrete Materials and Practice.* Edward Arnold Ltd., London, 1968. 4th edition, p. 398

KOLEK, J., An appreciation of the Schmidt rebound hammer, *Mag. Conc. Res.*, 1958, **10**(28), 27–36

ELVERY, R. H. AND FORRESTER, J. A., *Progress in Construction Science & Technology* Chapter 6, pp. 175–216, Medical & Technical Publishing Co. Ltd., Aylesbury, 1971

FORRESTER, J. A., *Gamma radiography of concrete.* Paper presented to Symposium on Non-Destructive Testing, London, June 1969, Cement & Concrete Association, p. 19

TOMSETT, H. N., *Ultra-sonic testing for large pours.* Paper at the Concrete Society Symposium on Large Pours, Birmingham, Sept. 1973, p. 4

TOMSETT, H. N., Site testing of concrete. *Brit. J. Non-Destructive Testing*, May 1976

TOMSETT, H. N., The non-destructive testing of floor slabs, *Concrete*, London (March 1974), pp. 41 and 42

JONES, R., *Non-Destructive Testing of Concrete*, Cambridge University Press, 1962, p. 102

LEVITT, M., 'The ISAT—a non-destructive test for the durability of concrete', *Brit. J. Non-destructive Testing*, **13**(4), (July 1971), pp. 106–11

APPENDIX 5

Notes on the Use of the Rapid Analysis Machine for the Determination of the Cement Content of Fresh Concrete

Until the introduction by the Cement and Concrete Association of their patented equipment for determining the cement content of fresh concrete, determinations of cement content had to be made by a chemical analysis of the hardened concrete. This meant that there could be no check on site at the time the ready-mixed concrete was delivered, nor when concrete came from the batching machine, on one of the most important aspects of a concrete mix, namely, the quantity of cement in it. The reasons for the need for an adequate cement content and a low w/c ratio in concrete in order that it should be durable and protect an embedded reinforcement have been discussed at length in various chapters in this book. For many years there has been a pressing need for the development of a quick and reliable method of estimating the cement content of fresh concrete.

The principle of this machine, which is known as a Rapid Analysis Machine, is that of separating material in the mix which has the same particle size as cement. This means that if there is other material in the mix which has the same or very similar particle size, for example, pfa, then the Rapid Analysis Machine would not be able to distinguish between the cement and the pfa. However, it is normal practice for specifications to include a requirement that no admixtures are permitted in a concrete mix without special approval being obtained from the professional firm who is responsible for the design and supervision of the project. It is reasonable to assume that reputable contractors and suppliers of ready-mixed concrete would not deliberately include in a concrete mix, material having the same particle size as cement without informing the client and obtaining his approval first. If this were done, then the firm responsible may find themselves in the position of being held responsible for the removal of a large part of a structure and its rebuilding and this would certainly not be worthwhile. Some sands contain silt, but an allowance for this can be readily made by a sieve analysis of a sample of the sand.

In a series of trials carried out in 1975 it was found that under practical conditions the machine tended to slightly overestimate the amount of cement in a mix compared with the declared weight of the cement by the suppliers of the concrete. However it would be reasonable to assume that the machine can estimate the cement content with a degree of accuracy of about ± 20 kg, or a range of ± 15 to ± 25 kg.

Index